A USER'S GUIDE TO OUR PRESENT WORLD

A USER'S GUIDE TO OUR PRESENT WORLD

What Everyone Should Know
about Religion and Science

HERB GRUNING

Foreword by Bruce Toombs

WIPF & STOCK · Eugene, Oregon

A USER'S GUIDE TO OUR PRESENT WORLD
What Everyone Should Know about Religion and Science

Copyright © 2021 Herb Gruning. All rights reserved. Except for brief quotations in critical publications or reviews, no part of this book may be reproduced in any manner without prior written permission from the publisher. Write: Permissions, Wipf and Stock Publishers, 199 W. 8th Ave., Suite 3, Eugene, OR 97401.

All Scripture quotations, unless otherwise indicated, are taken from the Holy Bible, New International Version®, NIV®. Copyright ©1973, 1978, 1984, 2011 by Biblica, Inc.® Used by permission of Zondervan. All rights reserved worldwide. www.zondervan.com The "NIV" and "New International Version" are trademarks registered in the United States Patent and Trademark Office by Biblica, Inc.®

Alternatively, scripture quotations marked (NRSV) are taken from the New Revised Standard Version Bible, copyright © 1989 National Council of the Churches of Christ in the United States of America. Used by permission. All rights reserved worldwide.

Wipf & Stock
An Imprint of Wipf and Stock Publishers
199 W. 8th Ave., Suite 3
Eugene, OR 97401

www.wipfandstock.com

PAPERBACK ISBN: 978-1-7252-9303-8
HARDCOVER ISBN: 978-1-7252-9302-1
EBOOK ISBN: 978-1-7252-9304-5

03/18/21

For my wife, Alice,
world's greatest life partner

CONTENTS

Foreword by Bruce Toombs   ix
Preface   xiii
Acknowledgment   xv
Introduction   xvii

PART 1: RECONNAISSANCE MISSION—SURVEYING THE TERRAIN

The Story Thus Far	3
Sounds Like a Plan	8
Long Overdue Comments	12
Answer to Joe	15
Alternative Models of Divinity	17
In the Twinkling of an Eye	26
Force of Habit	29
Who Made God?	31
Now That I Think of It	35
Harry (Hairy) the Troglodyte	38
Scientific Addenda	40
Religion-R-Us	43
23 and Who?	45
Identity and Me	46
Preliminary Conclusions	49

PART 2: LAYING THE GROUNDWORK

Text and the New Creation	57
Approaches to Scripture	59
Discrepancies and Inconsistencies	62
Foolishness to the Multitudes	67
The Jesus of the Text	69
Pet Peeve	74

Enter the Divine	76
Was Science Naturally Selected?	80
Cranial Signatures	84
The Presence of the Past	86
The Power of the Past	89
Let My People Choose	94
It Pains Me to Say This, but I Feel I Must	95
Changes of Mind	98

PART 3: CHARTING OUR COURSE

Who Is God?	105
"[S]o That God May Be All in All" (1 Corinthians 15:28)	108
It's Only a Matter of Time	110
What's a Body to Do?	115
For What It's Worth	119
Approached by an Angel	123
You Again!	126
Can There Be ESP?	130
Something Cannot Arise from Nothing	138
Just Can't Get Enough Science	141
Tentative Conclusions	145
Deliberate Conclusion	151

EPILOGUE

The Intrusion	157
Bibliography	163

FOREWORD

Reading a book by Herb Gruning can be a transgressive experience. Gruning's arguments don't stay in their lane. In universities, fields like theology, biology, sociology, philosophy, and physics have their own departments, their own canons, their own adherents. They also have, usually, fairly set ways of defining what is—and is not—an acceptable hypothesis or line of inquiry. We can study intuition, but not ESP. We can explore the idea of a gravitational field, but we're less sanguine about a morphogenetic one. Science has its heresies, of course, and in writing these pages about human nature, Gruning can't help but indulge in one or two. After all, it is a basic assumption of all his work that one cannot divide knowledge into rigid categories and insist that *this* can never be part of *that*. It just doesn't work that way. Nor should we want it to.

Gruning and I met as undergraduates and have been friends for most of our adult lives. He was one of the key players of my early twenties, in fact. We were young evangelicals, curious about the world, eager to reconcile faith and reason, and happy to discover that they were not such odd bedfellows, after all. Gruning was already bridging philosophy, science, and other fields when we met, and I loved philosophy because it meant I could study *anything* and call it research!

Regarding human nature, our explorations back then led us to reject body-soul dualism as both unsupported by Scripture and philosophically unsound. Gruning describes his memory of it in the first section of this book, and his account of it sounds very much like the conversations we were having at the time. By the end, as he puts it, he had "moved from body: bad; spirit: good, to both are good and united, a position that lasted for the bulk of my adult years. Until it eroded." The rest of the book can be seen as an account of that erosion and what was constructed on the newly cleared terrain.

Our experiences were different. I too suffered an erosion, but not of that particular belief, which I still think I was right to reject. But I went farther and, after decades of intellectual and ethical struggle (including a lot of, well, soul-searching), eventually abandoned not just dualism but Christianity altogether. This culminated in a sort of agnosticism I don't like to call by that name, since I think it implies a claim either that I have knowledge (that something is unknowable), or that I am happily ignorant (but I would rather believe), or that I am somehow proudly skeptical and hostile to my former religion. None of these is accurate in my case. But agnosticism is uncomfortable. The fact that *I* can't see the truth does not mean that there is none to see. I'm also not sure it helps me understand human nature. Now, it is true that a forty-year dialogue with my friend is likely influencing my response to his book. But still, much of what is said here, minus the theology (but not all of it, interestingly), seems very convincing to me. And where we do not agree (yet), I find Gruning's argument sufficiently challenging to force me to give a second look to its conclusions.

On the one hand, we are members of a particular kind of biological species living in a universe dominated by physical laws that we think we are beginning to understand. A large majority of us also thinks we are a particular kind of *meta*physical entity, though specifically what *kind* of entity has been the subject of considerable debate. Those who believe this also tend to believe that we live in a universe populated by other metaphysical entities. Furthermore, we are beings of rich subjectivity who are both creators and products of culture.

Thus, in seeking to understand what it means to be human, Gruning has turned his thought not only to the Bible and metaphysical speculation, but to physics, evolutionary biology, ethics, psychology, parapsychology, and of course theology. He is able to link authors like Alasdair MacIntyre, on the importance of narrative and metanarrative in the development of human selfhood, with scientific studies of brain plasticity and the radical Sartrean assertion that *choice* is the essence of humanity. He has thus taken seriously the argument that humans are in many and important ways agents capable of their own self-creation. Then he has invoked a critique of scientific bias that we have no choice but to take seriously, and brought in examples of interesting and potentially important brain and mind research, passing by way of a study of the Bible that is at once close and eclectic, only flinching a little bit as his argument draws him all the way, nearly inescapably, back to dualism. At every moment he is generous with allies and opponents alike, treating the entire subject with good humor and his trademark dry wit.

The book is challenging and sometimes difficult, but written in a way that should engage any interested reader. The short sections mean you can stop at almost any time, but the discussion carries you along so that you hardly notice. It is in many ways an eclectic read, and many of the questions raised remain unanswered. But this is to be expected: the study of human nature is the study of a multifaceted animal that cannot be understood without exploring vast universes, both within and without. This book is an excellent start.

Bruce Toombs
Humanities Department
Law and Civilization Program
Champlain College, St-Lambert, Quebec

PREFACE

There are times when we are thrust into the role of social critics. People have taken to the streets to express their grievances on a variety of issues. Some are incensed about illegitimate treatment at the hands of those in authority, the offended party usually a minority group and the reason often based on dermatological pigmentation. Others are acrimonious about lockdowns and the inability to live their lives and conduct business as they see fit, often for selfish reasons and with ostensible disregard for the common good. This does not speak well of human nature.

But these are not the only human endeavors to be socially critical about. Politics and political systems tend to be self-serving in ways more glaring than simply a failure of nerve. The two main enterprises most pertinent to the current work, though, are religion and science, for they also do not always deliver on their promises. Both have and have had their salutary periods, but also their detrimental ones. Unbridled science and technology have placed us at odds with the environment, while liberal religion ordinarily champions ecological causes and movements. Conservative religion sometimes stands in the way of concerns such as family planning, while science and technology could readily come to assistance. Can religion and science not combine their efforts?

As we find ourselves in a time when certainty about the future in general and ours in particular is lacking, the likelihood of asking not only temporary logistical but enduring fundamental questions regarding what human nature is like in the here and now, and what it might be like in the hereafter, is elevated, and arenas where answers are sought need to be broadened. Both religion and science can be brought to a similar deliberating venue, in the same room if not also around the same table. Perhaps there is even warrant for just such an association.

ACKNOWLEDGMENT

Many thanks to Kristine Jongepier for the preparation of Illustration 1.

INTRODUCTION

I suspected as much. I had originally intended to write a postscript to my previous volume for new thoughts which had come to light. As I labored, I came to realize that it became embarrassingly lengthy and I thus contemplated whether it should instead expand into another manuscript. The end result is the product you have before you—a series of afterthoughts that refused to quit. I discovered that each theme I addressed in what is now a post-postscript really deserved a proper treatment of its own. This I have endeavored to accomplish.

For the first time I am writing a sequel of sorts. More could be said about the topics covered in each of my volumes, but the previous one urged me to continue more so than the others. This means the current work also tackles human nature (specifically theological anthropology), most notably how we came by way of it and what is in store for it. These are my main concerns here. To achieve these ends, it seems appropriate to conscript the art of a short story in my tool kit.

When writers of fiction are asked why they employ it rather than simply saying what they need to say in a non-fiction format, they sometimes reply that they do not know how to say it adequately in any other way. I have come to realize that this is not a failing on their part, for at times theirs is the best way of putting it. Non-fiction may not successfully convey the heart of the matter. In the ordinary course of events, experiences lead to concepts formed to grasp them, which in turn lead to terminology used to capture them so that others can indirectly share in what was for the reporter a moving experience. This leaves the hearer or reader at a remove of at least two steps when interpretive measures are applied.

The initial encounter could very well contain a sizable portion of profundity and this impacts the experiencer subjectively. Once the event is reduced to concepts, rationality draws upon the familiar in a quest to situate it, and conventional frames of reference likely prove insufficient for an experience which could otherwise break barriers and shatter categories. Then in the next stage of objectification, language attempts to express what the concepts have already aimed at but missed. Hence I have found benefit in casting certain ideas in fiction alongside wrestling to unpack them in non-fiction. So I have elected to do both, trusting that the reader will also find the strategy salutary.

Fiction assists, I find, when I am at a loss for a particular kind of delivery, and in unwrapping the ideas contained therein I try to recover some of that loss. Besides, authors of fiction are allowed certain liberties that authors of non-fiction are not. I hope to seize upon that distinction. The intent in all of this is to share in the profound without losing its richness in the description of it. A futile attempt from the outset, but this ought not to deter us from the task. Ask the apostle Paul if he thought the effort worthwhile. His Damascus Road and Seventh Heaven experiences transformed both his life and, with time, much of the Western world. He sought to translate experience into concepts and then into language with a modicum of success, and we have inherited the fruits of his labor to our benefit.

PART 1

Reconnaissance Mission—
Surveying the Terrain

THE STORY THUS FAR

First by way of *apologia* or, dare I say it, memoirs (when is one too young to have them?), I offer the following. In my initial volume I strove to impress upon the process metaphysical and theological community that a merger between mathematician-philosopher Alfred North Whitehead's (to be outlined below) and physicist David Bohm's thought might best be served by placing Bohm as a foundation and building a Whiteheadian structure upon it, instead of the reverse. An unpopular innovation for Whiteheadians to be sure, for whom the idea was not met with enthusiasm. In my second work I attempted to convey the notion that Whitehead and Bohm are not the only schemes available in the metaphysical marketplace. Instead, there are several, through the efforts of those like James Lovelock, Thomas Becker, Rupert Sheldrake, and Pierre Teilhard de Chardin, among others. In my view, these remain live options as I compose these words.

In my third offering I considered two things. Whereas truth is not as obvious as previously supposed, the world is also a stranger place than originally anticipated. Anomalies abound. We might assume that, say, plants are relatively passive and inert when it comes to more active animal responses to their environment, but even flora react in ways suggestive of a central nervous system as found in fauna. Next I outlined that there are also alternatives when it comes to the concept of God. They lie on a spectrum between fully transcendent at one extreme to fully immanent at the other, while some combination of the two rest in the positions in between.

Lastly, my immediately previous work, after examining the nature of God in the third work, struggled with the topic of human nature. In it I investigated the notion that there are greater differences in degree than in kind between humans and other animals. Plus, what it means to be human includes the use and control of fire and engaging in artistic endeavors. Whales sing, bees dance, and beavers build, meaning music, choreography, and architecture find themselves in rudimentary fashion in the animal world (or better to say the other-than-human animal world). Yet beavers do not decorate the interior walls of their lodges with visual art, like portraits of forebears, nor does any other animal regale its community with tales about its ancestors or make myths about creatureliness.

Such activities are distinctly human and find no correlate anywhere else in the animal world. Also mentioned was the ambiguity of our natures. In addition to the ambiguity of our anthropic natures—we exhibit personhood but are also continuous with the animal world—there is the ambiguity of our characters. By this is meant our ethical attributes and the extent to which we display more virtues than vices. Even the best of us oscillate between the two and thus cannot be labelled as completely one or the other. At times we are saintly, at others we are offenders, meaning our allegiance is not always clear and we thereby betray our identity as double agents.

But the story begins much earlier. Consider the following a deposition of sorts. The culprit: misdirection. It all started when I was but a tender neophyte in socialized religion, specifically a Christian of a fundamentalist stripe. The assumption therein includes, among other things, notions about the makeup of humans, namely that we bear two or three components: a body and a spirit, sometimes with the latter being subdivided into mind/soul and spirit (the technical terms are dipartite and tripartite, respectively).

The soul/spirit is that which survives death and this makes the hereafter a blissful one unencumbered by physical limitations ranging from the ability to see in only one direction at a time, to the need for sleep, to backache, to cancer. Such a concept seemed innocuous enough, even reasonable. Though unsupported by science—a shortcoming hardly causing a ripple for my camp at the time—it reinforced the intuition that there is something more to us than meets the eye. What makes up the non-physical me outlasts the physical me. This is as it should be. God's got the matter (no pun intended) well in hand. "[T]he spirit returns to God who gave it" (Ecclesiastes 12:7).

Then came university, where all assumptions are up for review, save those which particularly benefit the university itself, such as the insistence that higher education is good for you, so you need it (together with the Enlightenment view, drawn from the Socratic dictum that the unexamined life is not worth living, and that faith gets in the way of reason, all the while not recognizing that these are faith statements in themselves). Here materialism reigns. There is no sense talking about what lies beyond the senses because that is non-sense. While this did not deter my extra-material leanings, it did leave them exposed. I responded to the threat in the following way.

There was a group of Reformed thinkers, Calvinists by another name, who came to the rescue. They offered both a diagnosis of and a prescription for my doctrinal malaise, though I did not realize that I had an ailment. They informed me that we have unwittingly espoused an ancient Greek, specifically Platonic, approach to understanding the human condition. To these ancient thinkers change implies corruptibility, so anything that undergoes any alteration, like our temporal bodies, which grow, develop, mature, decline, and perish, is devalued. On the other hand, that which remains static, like God and the eternal soul, are highly valued. They find their home elevated into a realm of Forms or Ideas, accessible only by a non-physical activity, namely contemplation, an exercise that philosophers, of course, can cultivate and refine.

Plato even went so far as to refer to the body as a prison house of the soul (definitely not a fan of the material world), since it hampers spiritual pursuits and becomes an impediment to the soul. All the while, we had been assuming that this doctrine known as dualism was precisely the biblical position, that the Bible's message fostered the perspective that the world was something to be escaped from in favor of a non-material heavenly existence. These Reformed types declared that this was a misreading of the Scriptures, that they did not in fact promote a dualistic picture. The battle line to be drawn is not between the flesh and the spirit, the first being carnal and displeasing to God and the latter lofty and pleasing, but between lives that are regenerate versus unregenerate. If we remain in the latter, we are degenerate; if instead we keep in step with God's Spirit, then we become "transformed by the renewing of our [hearts and] minds" (Romans 12:2).

God does not reject the body or the world but created them as very good, and the apostle Paul even refers to bodies sacrificed to God as holy (Romans 12:1). Admittedly they have become distorted, but God works to reverse the trend and re-establish their former goodness. God values the body and the world so much that God even raised Jesus bodily. Biblical enthusiasts sometimes forget that the love of God for the world made Jesus expendable (John 3:16). Resurrection then reflects the importance God places on the material world.

In fact, Revelation 21 announces that there will be a new heaven and Earth, upon which God's followers will dwell, as opposed to our being whisked away from the Earth into heaven. So we had better get comfortable with the idea of physicality, since we are going to find ourselves in it, in some description, for the long haul. God made us psychosomatic unities— souls/minds inextricably tied to bodies. This sounded like a radical direction for me to take, so I moved from body: bad; spirit: good, to both are good and united, a position that lasted for the bulk of my adult years. Until it eroded.

Subsequent excursions into the Bible continued to yield fresh insights. Contrary to the notion that the New Testament and Paul in particular did not present a dualistic portrayal of God's economy, time and again I came across passages which resisted such inclusion. Statements from the Jesus of John's gospel, specifically, "The Spirit gives life; the flesh counts for nothing" (6:63) (at least in comparison to the Spirit), were a case in point and gave me pause. Akin to Thomas Kuhn's exaggerated view of the "structure of scientific revolutions," I noticed biblical anomalies that would not surrender to the non-dualistic paradigm, till the point where I reached a critical mass of them, when confidence in the paradigm waned, and I suffered a theoretical/theological breakdown. I could support the strategy no longer and sought a replacement paradigm.

Curiously, this revolution yielded not a new paradigm but a reformulation of the initial unexamined one. The question turned on the type of dualism at issue. At many an instance Paul wrote about dualistic themes that I could no longer ignore. He estimated, for example, that it was salutary to depart and be with the Lord, but concluded that it would be more beneficial to remain in his ministry for the sake of his flock (Philippians 1:21–24; 2 Corinthians 5:6–9). This implies that there is something about us which departs the body at death and survives to be ushered into God's presence. A definite dualistic posture. Further, Paul, in the capacity of tentmaker, describes the body not as a prison house of the soul, as does Plato, but as a tent that we reside in and which is to be removed at the time of death in favor of another—this time eternal—dwelling (2 Corinthians 5:1–4). Besides Paul, the unknown author of 2 Peter echoes the same sentiment in an offhand way in that he claims to "live in the tent of this body" and "will soon put it aside" (1:13–14).

What is more, there will be a reckoning for "the deeds [we have] done while in the body" (2 Corinthians 5:10), suggesting that our enduring selves are distinct from and in addition to our bodies. There is, as such, an "I" which resides in the body. This was for me the point where the Reformed view derailed and a second revolution took place. The previous paradigm became supplanted by the following. It occurred to me that the body need not be valued negatively. While true that it becomes discarded upon our passing, that by itself does not make it valueless. It can still be upheld as a good creation and retain its status as valuable, and we can continue to hold on to the idea of residence on a renewed Earth.

Even in our own development, we shed physical aspects once they have outlived their usefulness, such as webbed feet and hands while we are in the womb. They can safely be set aside, a process known as cell death, once our growth in the womb reaches a certain stage. We do not look upon webbing as something to be avoided and breathe a sigh of relief once it subsides (although there are some for whom webbing continues on into adulthood). Instead, there comes a time when their shelf life is reached and their expiration date comes due. But that does not mean they were always and inherently a setback.

There are several things we can experience bodily that apparently spirits cannot, such as the warmth of the sun, the embrace of an intimate partner, and even playing throw and catch. In fact, perhaps there are times when spirits envy the embodied. And on this topic of embodiment, we may consider the concept of angelic beings. When attempting to imagine what resurrection bodies might look like, angels come to mind. They are not the same thing, but both could be described as bearing spiritual bodies. Paul proclaims in 1 Corinthians 15 that flesh and blood cannot inherit the kingdom of God and that the corruptible/perishable must put on the incorruptible/imperishable.

There may be a parallel here with angels. Both they and we are creatures, in that we and they at one time were not and then came into existence at a later time. With God as the lone exception, we all had a beginning. As creatures, angels have some type of being within certain parameters. They are not omnipresent, as God is, but like us have coordinates or occupy a certain locale at one time and not another. They have location. As a result, they must have some type of makeup or constitution which takes up a location. Same with resurrected bodies. They and we can be identified by certain features, not least of which is our names. Angels are a different spiritual species, but their spiritual "bodies" might very well be analogous to what ours will be like (Luke 20:36).

All of the foregoing implies that not all dualism need be Platonic. God still values the world and the organisms in its biosphere, even though ours will become obsolete and require refashioning. Analogously, a similar thing occurs with technology. We do not use floppy disks anymore, but this does not mean they were never useful, though now they are no longer required. Computer enthusiasts needed them at one point. They are now devalued due to their obsolescence, but not because they can be placed in a category labelled "bad." Resurrection bodies are new and improved but they are not the first things of value. Value has preceded them. Floppy disks had their day in the sun and were handy for their purpose. But they have been superseded. The value judgment to be applied here is useful versus useless, not good versus bad. The latter is dualistic; the former is not. The Platonic variety has been replaced.

Despite two misdirections, I now find myself in the non-Platonic dualistic camp, so I am a dualist of a different flavor. Perhaps I should call myself a neo-dualist.

But enough about me; the time has come to fictionalize. What comes subsequent to it will be an unpacking of certain aspects of the short story as well as the above.

SOUNDS LIKE A PLAN

Sound is a precious thing. It is not plentiful in the cosmos. Sights you can get almost anywhere throughout the universe, but sound requires an atmosphere or some denser medium than deep space can supply. In these terms, ears should be more valuable than eyes, though in our lives a premium is usually placed on sight. In the grand scheme of things, however, sound is the rarity.

The man seated next to me was providing both sight and sound. This hospital ward also did both, but my room had grown oddly quiet. Perhaps it was the time of night. The man, whom I did not recognize, informed me that I had been brought in the previous day but was only now sufficiently lucid so as to hold a conversation. He told me that I had been wounded in battle but that I was convalescing nicely in this army hospital, though my full recovery was still in the offing.

"Any idea when that might be?" I inquired, assuming that he was one of the medical staff or at least a chaplain.

"I cannot tell," he replied, "but you must be mistaking me for a health care professional."

"So you're not?" I asked.

"Not at all," was his answer, "just an interested party."

"What makes you so interested in my recuperation then?"

"Call it an investment in the big picture. With our expanded view we see more clearly."

"How large?" I wanted to know. With my newfound vigor, I felt a little feisty.

"It's more longitudinal than latitudinal in nature. Actually, it's difficult to explain."

"Try me," I demanded.

"Well, it's not so much how you fit into the current frame, but the span of several lifetimes."

Puzzled, I asked, "Are you referring to my ancestors?"

"No, not to those who came before you, but to the you that came before you."

My response was measured. "That's crazy! You're right, I don't understand." And with that he attempted to explain to me in language I could follow but not comprehend that not only had my life been marked by military service but also of previous lives and battles, with me surprising myself that I would so readily play along.

He claimed that I had perished on the battlefield in the Franco-Prussian War; suffered the same fate as a thirty-year-old in the Boer War; had died yet again as a seventeen-year old, having lied about my age, in the trenches of Belgium; and then had fallen on the beaches of Normandy. Quite a resume, I thought.

"Your nationality alternated between German and British on those occasions." I had been in the military in one form or another, apparently, ever since. I would not have believed him were it not for the photos he produced. There were successive portraits in which I had a similar physique and facial features.

"And that's not all," he exclaimed, "look at the other figure in each instance." He was pointing to the person who he informed me was supposedly my spouse. "Notice how all of them are black-haired, brown-eyed, and of approximately equal height. You have a similar taste for type in each case."

"But how do you know that's me?" I asked, becoming increasingly uncomfortable.

"Have a closer look," he insisted. "The birthmark on your hand which you currently possess is also visible on each photo."

Wanting to change the subject, I asked, "How did you come by way of these photos? Are you a historian?"

"I am definitely interested in both your and overall history."

"Why me?"

"Each of us has his assignment. I say 'his' although we do not have gender as such."

"Who is us?"

"We are called upon to assist in history."

"In what way?"

"Call us facilitators."

"What do you facilitate?," my replies becoming increasingly tempered.

"You have a lot of questions. That is understandable. We encourage the searching out of the particulars."

"How many of you are there?"

"As many as are required for the tasks that emerge."

"So who took these pictures?"

"Ordinary photographers; we do not operate cameras."

"How did they come to be in your possession when they are someone else's property?"

"Our executive branch will duplicate what is most needed. We have a policy of non-interference unless given clearance to do so."

"You can do that?"

"Remember the tale of the loaves and fishes?"

"I guess you are not in any of these shots?"

"No image of us can be recorded through the usual channels."

"How about the unusual?"

"Which no one is privy to."

"You have covered all the bases, haven't you?"

"We seek to be thorough when given leave."

"Do all of you look the same and speak English?"

"We come in various forms and communicate in ways that can be understood."

"What would be the point otherwise, presumably?"

"Quite so, but let us get back to your case in particular."

"I know of a military family in which the grandfather fought in Korea, the father in Vietnam, and the son in the Gulf War. Is this the kind of task you are referring to?"

"Once again, those are lines of descent, not as in your case previous lives. But yes, they too receive attention when warranted."

"And who decides what is warranted? Who is in charge?"

"It works by way of decision by committee."

"So you are one of the agents."

"More or less. We have a certain amount of latitude when enabling certain courses of action."

"So let's get this straight. You here seated next to me, uninvited, are neither medical staff nor chaplain. Are you even human?"

"That would not be an accurate description."

"So your form does not reflect who or what you are."

"I carry no ID card, for none could adequately identify me. I have neither ancestors nor descendants." It's like he knew what I was thinking.

"Do you often make these appearances?"

"We keep them to a minimum. I will depart before the next staff person makes his or her rounds."

"So no one would believe me if I were to tell them about this experience."

"Would you?"

"Not likely. So how did I rate this attention when my forebears—excuse me, previous selves—did not?"

"Recall I mentioned latitude. We could not bear the thought of you experiencing another death on the battlefield. So you were mostly shielded from the spray of bullets and shrapnel."

"But there is an afterlife beyond these lives, right?"

"That comes later. First let's concentrate on this life."

Still not sure that I could accept these terms and conditions, I must admit that I found it strangely comforting that someone or something was in my corner, even though I did not have complete control of my circumstances. Yet I could not help but feel indignant about the seeming injustice of it all.

"Why wait until this time around?" I insisted on knowing. "And why not for some of the others in the trenches?"

"You are not the only one. Besides, it also has to do with order, which is enhanced by our course of action and would otherwise be diminished beneath an allowable minimum."

"So it's all about order?" I asked in my bewilderment.

"That's not all. The wife in your previous life also made the trip with you into your present life. The love you shared transcended the merely earthly plane and struck a chord with us as well. We could not help but grant it another go-round."

"So you adore a good love story! And my current wife is, or was, also my previous wife?"

"Love of this expansive quality cannot be limited by human life spans; they erupt into new lives."

And with that a sound came from the corridor. I turned to look and it was the orderly who came in to clean up. I turned back only to find an empty seat beside me. Was I just dreaming? Was it just the medication talking? It occurred to me that I never asked his name, if he even had one. The experience left me with more questions than answers.

I asked the orderly if he saw a man leaving as he was entering. I tried to describe him to the best of my ability.

"No one matching that description," was his response. "Besides, there is no one on staff like that and visiting hours are over."

Just then the physician on duty dropped by to announce that my health was improving. I liked the sound of that.

LONG OVERDUE COMMENTS

Having one's words quoted can be an affirming experience, even a flattering one. Unless, of course, it becomes a back-handed compliment. I mentioned in the above my first volume and now I wish to present an instance in which this writing has been referenced. The respondent, whose name I could not track down (the particulars were never found—evidence that Google is not omniscient), in this case takes note of my offering disapprovingly. The source of the disquiet is my methodology.

The interest on the part of the respondent converges with mine in that the subject matter in the objector's crosshairs is divine action. The worry is that my tactic of treating this topic without initially establishing the existence of a divinity to speak of amounts to an unwarranted hastiness. One must begin, in the critic's view, with the foundation of God's existence and supply reasons for it before one can legitimately advance to the second stage of what it is that an alleged God allegedly does. The claim that to do otherwise is unduly skipping ahead. More to the point, beginning where one should end up is question-begging, surreptitious, and dishonest. I must now answer these charges.

My initial counterresponse, I suppose, is that it would be underhanded if I did not reveal at the outset how I was going to proceed. It seems that I have indeed covered that ground by stating the necessary qualifier. Next, what to the objector is not an honest option is for me a functional one, since it bypasses an issue which can never be settled, so as to move ahead to one about which more headway can be made and without which the second theme could not be addressed by the rules the critic adheres to. If we were insistent on resolving the first issue before tackling the second, then not even science could get very far. We know neither, strictly speaking, what matter is nor the components of what is to us an invisible universe, such as dark matter and dark energy. Nevertheless, science goes about its business. Would critics suggest that this is a failing on science's part?

Besides, my approach enables more inquisitors to be brought to the table. Atheists are welcome to contribute, once the divisive concern of God's existence is bracketed for the moment, long enough for a fresher approach in religion and science to be undertaken. If it can be argued successfully that other versions or models of divinity, for example, can accommodate more of the natural world in which theists and atheists alike find themselves, then they might recognize that what were once thought of as obstacles to belief, such as specifically classical renditions of divinity, which certain theists are equally loathe to defend, need not remain roadblocks anymore. Perhaps other approaches to divinity, like process thought, would not hold the same disapproval rating for the atheist crowd.

If so, then the alternative to the objector's initial concern can receive a substantial hearing. Better in my view to open the door to discussion where all parties can participate than to hammer at the previous door with diminished prospect of its being opened. It becomes a matter of strategy, and neither humans nor the deity always use the same one. My having addressed the volume to an audience including atheists who have moved beyond the traditional God scheme is not something about which I apologize.

Other points of note: the critic's mention of those intellectual luminaries who are nevertheless believers is an appeal to authority. The intelligent can be incorrect as well. Moreover, admitting that the existence of God debate can be ultimately disillusioning does not bode well for theists. There is less motivation for wrestling with an opponent if s/he is disinterested in interacting by reason of the futility of the pursuit. And finally, the conclusion which the objector reaches, that establishing the probability or likelihood that a deity exists is sufficient grounds for operating as though this were the case, since this view is already adopted by many, such as scientists working with uncertain theories, could be unconvincing to the target audience, for uncertainty about how to interpret the quantum world, using one of the respondent's examples, is one that they and we can live with. It happens all the time and life goes on.

Yet whether God exists has much more gravitas. It is a major query which one would assuredly like to get right (aside from Pascal and his wager). One can afford to be inaccurate about some aspects of quantum and still reside nicely in the current natural world, whereas no one would likely relish being on the wrong end of the God debate in anticipation of leaving this world. And note that not every theologian is a theist, nor is every thinker who engages in the question of divine action.

Parenthetically, Blaise Pascal was a sixteenth-century French philosopher-mathematician who submitted the following argument: since one is unable to prove neither the existence nor non-existence of God definitively, it is best to hedge one's bets on God's existence and believe, for the stakes are higher if God exists and one does not believe than if God does not exist and one does believe. In the latter case, no harm, no foul. The wager is, obviously, ultimately a non-committal position, for one accepts the terms only in the face of the perils involved, not because one's heart is actually in it. From God's perspective, one must ask, is someone who believes simply to avoid danger the kind of follower God would seek to invest in?

Having said all of this, I need now to address what follows.

ANSWER TO JOE

I must mention that there is admittedly one item left unresolved in my first book, namely, in the introduction, how one would respond to Joe Carter's claim about divine activity in terms of his famous home run belt to win the second consecutive World Series title for the Toronto Blue Jays baseball club in 1993. I never answered the question I posed at the outset. I want now to remedy this.

Joe's belief is that his classical version of God was involved largely during the flight of the ball, straightening out its trajectory so that it would remain in fair territory and count as a home run. My thinking is that if the divine were to have been factorial in this event (and in Whitehead's view that is a requirement—more on this shortly) then it would have occurred prior to the flight of the ball. God could have impressed upon Joe's thoughts the idea of swinging the bat a certain way or to anticipate the pitch he was about to be served.

Alternatively, God could have brought to Carter's recollection the scouting reports on this pitcher and the next pitch he was likely to deliver. Or there could have been a combination of these approaches. In any case, it seems to me that divine activity would occur either at or prior to the point at which the bat makes contact with the ball—whether in the prediction of the following pitch to be thrown or in the mechanics of the bat swing or both. Subsequent to that point, it appears that ordinary physics would take over (something about which the process God also has an intention).

The God of the biblical text, however, is acquainted both with standard physics as well as psychology. A divine influence appears in dreams and visions together with intervention in nature and history, as we are informed in passages such as Psalm 107:29, where God "stilled the storm to a whisper; the waves of the sea were hushed." This agency is what certain persons find irksome with the traditional theistic view, since this betokens a deity that operates by way of miracle. The world of physics is somehow flawed, according to their reasoning, and requires God's intermittent attention, since the world by itself under its own steam could not accomplish either its or God's will.

For this divinity, adjustments are called for whenever God's intentions are in jeopardy, so as to ensure that the divine will is secured, at least for the time being, until the next flaw in the system arises. Every time an issue surfaces, this divinity steps in to correct the mechanism, prompting questions on the part of some, like what is wrong with the creation such that God needs to step in periodically? The world could not have been created perfectly, for if it were it would not require God's tinkering, or, alternatively, if it was created perfectly, it is not perfect now. Neither option is attractive.

Critics of the traditional understanding object to the strategy as being unworthy of the divine economy. For them, a God that (or who) functions as a fine-tuner is unbecoming of their vision as to what a deity should be like and the role a proper divinity should assume. Compounding the problem is the seeming arbitrariness of those situations which call for the divine attention and those which escape it or, what is worse, do not deserve it. These critics have a point.

Thankfully, this is not the only rendition of divinity on offer. The others include the process and openness views. The former overcomes both problems and to some is impressive for its explanatory power; the latter, regrettably, does not successfully address concerns of arbitrariness. To chart the differences, the degree of authoritarian rule God either can or does exercise decreases from traditional/classical to openness to process approaches. More on these themes presently.

Accordingly, the answer to Joe is that his view draws on the interventionist scheme, but that is not the only candidate. If God is of the openness type, then why did God favor the Blue Jays and not the Philadelphia Phillies? Were there either more or stronger devotees on the Toronto team or its fans praying for a favorable result? Or if God goes by the process format, then God could only operate psychologically and viscerally. While this divinity draws nature to behave orderly, according to what science would describe as physical laws, there is not even the possibility left open to God to adopt interventionist tactics. Joe Carter's God is the traditional one whom, say, Thomas Aquinas would endorse. Yet if God were to be of another kind, then the answer would not be so straightforward. Whatever might be the most accurate account—and here I betray my own loyalties—thank God, and way to go, Joe.

ALTERNATIVE MODELS OF DIVINITY

Since you can't tell the players without a program, it seems best to outline in brief what certain figures contribute to our discussion so as to set the stage. I consider each to be significant if for no other reason than they have been important for me—factorial to my own worldview development. Some I take with more seriousness than others, but each supplies ideas which have inspired me to pursue my own lines of inquiry. We will focus our attention on two main ones and begin with Whitehead. Following this, we will chart the distinct features between the three.

Whitehead was driven to sculpt a new vision of divinity based on his personal experience of learning of his son's death in the Great War, later renamed World War I. This ignited a resistance on his part to the classical rendition of divinity, championed by luminaries such as Augustine and the aforementioned Thomas Aquinas, which maintained a sovereign deity having authoritative rule over all of history in general and in particular. If God enjoyed all the power and foresaw all that was to occur, which could not be altered since otherwise God could not have foreseen it, then creatures do not possess free will and God becomes the puppet master who holds all the strings or the dealer who holds all the cards. Should this be the case, then no occurrence falls outside of God's purview, "for who can resist [God's] will?" (Romans 9:19), which Paul the apostle implies is a foolish question.

A divinity of authoritarian rule apparently has the right to do as God sees fit with what God has made. This being the case, Whitehead reckoned, meant that God killed his son, for what other conclusion could one draw? In fairness, it must be stated, not all of Christian theology fits into this traditionalist mold, for it is of a specific stripe known as Calvinistic, from one of the key figures of the Reformation—John Calvin. Not all Christians or theologians follow Calvin in this, although, as we will determine (pun intended), it is theoretically difficult to avoid.

Whitehead was disgruntled with this portrayal of deity and so decided to draft one of his own. His impetus and what spurred him on to a different model was the problem of evil or the theodicy problem—a justification of the ways of God. He decided that if we live in a world bereft of free will, making everything foreordained or predetermined, then all that happens bears God's approbation. His counter-conception is referred to as process thought. There are other process formulations, but his is the most systematic approach.

Whitehead commences by insisting that what is truly real in the cosmos is not substance but experience, and in so doing turns science on its head. The stuff of the universe is experiential in nature, a category which he takes metaphysically seriously, in opposition to those who claim that the foundation of the universe is substantial matter, which Whitehead urges is actually insubstantial. To think otherwise is to commit what he terms the fallacy of misplaced concreteness. He attempts to explain this in the following way.

As an initial statement which will require unravelling, experiences essentially come together and are synthesized according to a plan and then produce an object which is relegated to the past. This is accurate in reference to Albert Einstein's relativity theory, for we cannot receive information about an observed object faster than the velocity of light, thus any information we obtain is how the item of our concern appeared in the immediate or distant past, depending on how far removed the entity lies from us. The farther away, the more the information comes from the remote past. Hence all objects are in the past, some more so than others.

For Whitehead, entities have a subjective phase, in which experiences gleaned from our data world of experience rally around an organizing center, to the extent that the entity in question boasts one, which shapes the experiences according to their own vision for themselves and their future. Since we cannot reach the objects in the past, given that we do not reside in a world of external relations in which substances merely collide with each other, there is an internal relatedness to them, where we draw upon the data embedded in them which mattered to us or which we felt strongly about. The term *feeling* here is used philosophically and is intended to convey what draws us to response. All entities feel the data afforded by the objects in their past and respond to them. We select the data we are exposed to and integrate them so that they conform to our own purposes.

Once the responses are complete, which can be of the order of microseconds for what was previously thought to be inanimate objects to tenths of seconds for humans, the formative stages yield an object, aptly named the objective phase. From the past they influence the present and to the past they return as one more installment of objects. Entities become and lead to being, but only for a moment. They then supply data for the next round of becoming. The reality about the world is that it is processive. Process predominates and is preeminent over being, and the process of development from subject to object constitutes one event. With the passage of every event is another data set that nascent entities can draw upon, ad infinitum. Counterintuitively, flux and not the static is the basis of reality.

How then does God fit in? On the one hand, speaking of order, God is its source, base, or foundation. That is, for entities which enjoy little in the way of free will or choice, which Whitehead refers to as self-determining power, the task set for them is to repeat the past. This provides structure and the type of order that science investigates, a reliable kind of consistency that can be depended upon. It is important to note that all entities appear somewhere on a hierarchical scale of sophistication in terms of their capacity to integrate the information from their data world of experience, from the neutrino to the numinous, from the electron to *Elohim* (one of the names of the Hebrew divinity), from the gadfly to God. Nothing escapes or is left out of the operation of this process; it is all-inclusive, all-pervading, and all-encompassing.

On the other hand, God is also the foundation of novelty, meaning that experiences can be combined to produce the genuinely new. The greater the self-determining power an entity possesses, the greater the potential for novelty. Contrary to what the teacher in Ecclesiastes bemoans, that there is nothing new under the sun (1:9), the creative advance toward novelty is never-ending, since it cannot reach a maximum. (I suppose there could be such a maximum were the universe to end in a frozen death where there would no longer be any kinetic energy available to do work. This is only one, though the most likely, scenario.)

Here we may consult Aristotle's four causes. He suggested that causes have four types. Employing an architectural analogy, a material cause is that which goes into the construction of a building, namely wood, steel, concrete, nails, and so on; an efficient cause would be the laborers who utilize the material causes in their work, wielding their hammers and saws as they proceed; a formal cause would be the blueprint, design, pattern, or plan according to which the laborers apply their efforts; and lastly the final cause is the (alphabetically) aim, end, goal, or purpose for the structure itself and for which those involved go to all the trouble, specifically shelter from the elements and so forth.

Modern science has excused itself from the need to make reference to the latter two causes, particularly because designs and purposes do not lend themselves to testing in controlled laboratory conditions, and so do not fit into the scientific regime, leaving science to concern itself only with the former two. We have Galileo to thank for this trend toward a methodological change in regime. In Whitehead's view, however, the latter two can be reinstated, since entities on the upper end of the hierarchical scale can exercise them. (Feminists need not bristle when they encounter this hierarchy, for it is not a male-dominant one. Humans appear at one rung on the ladder, both men and women.) Both God and humans harbor plans according to which they direct their efforts and initiatives.

Speaking of initiatives, God both initiates and is subject to process. As for the former, God keeps the process as a going concern by placing an entity in a location where it may commence its next round of becoming, essentially by giving it an address. God further offers the entity-that-is-about-to-develop God's own purpose for it as one datum or data set, which God hopes the entity will grasp and take upon itself to integrate into its own purpose for itself. This is a small yet potent offering since it appeals to the entity's better nature, in that it involves some combination of virtues, such as truth, beauty, goodness, and other abstract ideals found in, say, divine moral attributes, which God hopes the entity will make concrete.

God knows the best for each specific entity at any given time, due to God's wide-ranging vision, and seeks to impress this upon the entity. Despite its being but one single data set amongst the welter or barrage of influences an entity is bombarded with, that by itself does not drown it out. Scriptural support for this could be taken from Jesus' parables of the mustard seed, in which a small seed grows into a large plant, and the yeast, where just a pinch works its way through the entire dough (Matthew 13:31–33).

God's purposes might be the most irresistible force in the universe, though, given the self-determining power of higher entities, it is not so irresistible that it cannot be resisted. We have the ability to trump God's leading, urging, and what Whitehead calls a "lure for feeling," so as not to heed God's call. God's plan is always to increase the creative advance toward novelty in the world, and any time this program is rejected, an element of disorder is injected into the world, another term for which is *evil*. And as already intimated, God is also subject to the very same process which God sustains. Even the entity at the top of the scale is not excused from process. This God also becomes or evolves.

One way in which God changes is in the content of God's knowledge. God comes to learn what each entity will do with God's purposes each time an entity experiences a round of becoming. This marks a moment of discovery for God. God might very well be the best prognosticator in the universe, but this still does not amount to certainty.

Furthermore, God includes the world—the world is part of God. God is much more than the world, for God is mostly an inexhaustible reservoir of potentialities containing the abstract ideals just referred to. This is God's eternal essence, analogous to God's mind. The other aspect of God is the world of all other entities, cosmos-wide, known as God's concrete actuality, analogous to God's body. As the world changes, so too does God, entailing that what entities do with God's hopes are inherited into God's own makeup.

Incorporating traditional categories into the process strategy, we are co-creators with God, and in presenting entities with a purpose to rally their efforts around, God becomes creator. In upholding the process by unceasingly urging entities on toward greater novelty, God becomes sustainer. Finally, in inheriting what the world does with God's purposes, forever imprinting them into God's own becoming, God in turn becomes redeemer.

A comparison of divine attributes is found in the table below, but first we need to introduce a third party. The openness model, also known as free will theism, spearheaded by the late Clark Pinnock, resides somewhere in between the two extremes of classical and process theism. The similarities and differences can be described using Charles Hartshorne's and William Reese's classification scheme of five letters, in which E stands for eternity, T for temporality, C for consciousness, K for knowledge of the world, and W for world-internality, that is, God as containing the world. Here are some of the combinations.

Simply being an E is the God of Plotinus, in which God is analogously seen as a volcano who spews out from the peak that which increasingly becomes material the farther down the slope it descends. Toward the top can be found mind or soul, for instance, and at the bottom the extent of the world's material things. This God can boast merely bare eternity, conscious neither of itself nor that which emanates from it. Plotinus' position has the advantage, as some would see it, over Plato's in that material reality is good since it all stems from God, whereas for Plato matter is devalued since it changes, and only the static is highly valued. Change implies corruptibility, as Plato and other ancient Greek philosophers contend; therefore God must be static.

The next combination in view is EC, which is the God of Aristotle, who is conscious only of itself as the one thing worth contemplating. This God is not even aware of the world but engages only in divine navel-gazing. The combination that follows is ECK, in which God also knows the world and goes by the name of the classical deity championed by figures such as the Reformers Martin Luther and John Calvin, who drew from Aquinas and Augustine, respectively. Not until we arrive at the openness model, which Hartshorne and Reese do not include because the position was not as yet on offer at the point of their publication, can we introduce a T, since the God of free will theism is not outside of time but in it, and in turn is affected by the temporal process, the passage of time. As in process, the openness God does not know the future with certainty, but unlike process alters its hands-on tactics as the situation warrants and stops short of incorporating a W, meaning the world is not internal to this divinity.

Lastly, only in process do we find a W and thus a full complement of all five letters. In fact, exponents of process argue that any combination not including all of ECKTW is deficient. To contain all five letters, in their understanding, avoids the disadvantages of the other combinations and bears all the advantages the other approaches cannot manage on their own. As authors often do, they leave what they consider the best alternative until the end, and Hartshorne and Reese are no different. We are now ready to unpack the following table.

Table 1: Comparison of metaphysical attributes in classical, openness, and process divinities

	Classical (ECK)	Openness (ECKT)	Process (ECKTW)
Eternality	God has neither beginning nor end but is timeless, unaffected by the passage of time.	God is in time and is affected by the temporal process.	One aspect of God is an eternal essence, God's primordial nature, which can be likened to a mind.
Spirituality	God is non-corporeal and has no physical parts.	Same as for classical	The other aspect of God is the world, God's consequent nature, known as God's concrete actuality, which can be likened to a body.
Simplicity	God is a unity and is not made up of parts; God is not a compound.	Same as for classical	The world, as God's body, is made up of many parts.
Immutability	God is unchanging in terms of God's nature and purposes.	God's tactics may change depending on what humans do with their free will.	God evolves along with the world and so changes in God's concrete actuality.

	Classical	Openness	Process
Impassibility	God is passionless toward the world; God is unmoved emotionally toward it, since to be so moved would entail a changing God.	God cares enough about the world to seek to redeem it (John 3:16) and sometimes alters tactics so as to achieve this end.	God is moved emotionally toward the world and is, according to Whitehead, "the fellow-sufferer who understands."[1]
Independence	God is self-sufficient and does not require the world in any way.	God is dependent in so far as God alters tactics when the situation warrants, when humans use their free will in unplanned-for ways.	God is dependent in the aspect of concrete actuality, since the world makes up God's body.
Omnipresence	There is nowhere that God is not.	Same as for classical	God is everywhere that there is an entity, for these are the only places as such in the universe.
Omnipotence	God is sovereign and operates in terms of authoritarian rule, meaning the presence of evil is entirely God's responsibility.	God relinquishes sufficient power so as to enable creatures to exercise some measure of it, yet somehow retains all of it; meaning both God and humans are partly to blame for evil, leaving God open to the charge of arbitrariness.	God is not in a position to coerce, but can only persuade by means of luring entities toward feeling and grasping God's purposes.
Omniscience	God knows all things past, present, and future, as well as possible and actual.	God does not know the future with certainty, since creatures can exercise free will.	God grows in the content of God's knowledge as the free choices of entities come to light.

By way of commentary, it might initially seem that openness is closer to classical than to process, since we have identified as much in fully one-third of the attributes listed above. Yet the first and last items comprise a sea change, for some properties are more telling than others. For God to be

1. Whitehead, *Process and Reality*, 351.

temporal and not know the future with certainty makes the openness deity look more like the process God. What prevents openness from holding full membership in the process fraternity is the absence, even avoidance, of a W—the concept of a world internal to God. That marks an enormous difference. Curiously, proponents of the free will theism model have a more difficult time making compatriots in the classical camp than in the process. Perhaps classical regards both of the others as rebels from the lone orthodox fold and, via guilt by association, openness is equally as radical and hence heretical. Regardless of the labels that tend to be tossed about, all three options consider themselves as the correct one, while classical sees itself as the lone legitimate stance. I lament this state of affairs because I once belonged to the latter.

The difficulty boils down to this: the more power God wields, the less humans have, meaning God holds more responsibility for the presence of evil in the world. For the classical crowd, God enjoys all the power, hence, unrecognized by many of them, God has all the responsibility. For the openness camp, God extends some power to humans, entailing that both share partial blame, though, mystifyingly, God nevertheless retains all the power (?!). For the process folk, God's power is limited to persuasion, implying that virtually all the responsibility rests with humans.

The process version appears to be a tale of two divinities in terms of the dual aspect nature of God. The deity's "mind" overlaps with the classical God in that both are eternal, spiritual, immutable, and independent. But process gets to have it both ways because God's "body" makes up for the former, thereby disengaging it from the classical scheme. The strategy here in God's body is that God is still eternal, since there must always be a universe of some description, which may or may not accord with future astrophysical and cosmological findings, and is therefore also corporeal, mutable, and dependent. This is either pure genius or hedging one's bets.

A conversation could run something like this:

> "Is your God affected by what we do?"
> "Oh yes, God is definitely dependent upon the shape the world takes."
> "Oh good, so your God is not the type that is changeless?"
> "Well sure, but changeless is that part of God analogous to a mind."
> "That's certainly covering all the bases, isn't it?"
> "Yup, we've taken all the angles into account."
> "Doesn't that leave God a little thin when stretched so far, being well-nigh omni-flexible?"
> "Well, one person's thin is another's robust."

I will have more to say about my misgivings concerning the process scheme toward the end of the investigation in Part 2. For now, it seems openness would be a commendable middle ground were it not for the arbitrariness with which certain circumstances earn God's attention and which do not. For instance, why did tragedies like Katrina or Columbine not make the cut? Regardless, their step of bringing God into time modifies several things. Classical proponents dislike the move, for it humanizes God too much, and even more so in process (given that classical is always on heresy watch). Neither openness nor process views this as negative; rather, it becomes an advantage when attempting to resolve how a transcendent being can also be immanent in the world and relate to it. We can see that a single change can have wide-ranging ramifications.

IN THE TWINKLING OF AN EYE

Prompted by the Pauline phraseology in 1 Corinthians 15:52, where one type of body is exchanged for another (resurrection) and the time frame in which it occurs, I venture here into an epistemological and not just a metaphysical point, namely, Whitehead's understanding of the workings of knowledge, specifically decision-making processes. For Whitehead, these events traverse a number of stages from unrefined, bulk, bare-bones felt awareness to refined, considered, motive-driven response. These occur in an instant, though process thinkers refer to them as the quanta of existence having "temporal thickness,"[2] or, to coin a term, extended elementals, but not so quick that they cannot be unpacked. Take for example something we have all experienced and let us go through the various thought processes which combine to comprise a cognitive and visceral event.

Say you have grabbed a handful of trail mix and a certain recalcitrant raisin within it has jumped ship. It first lands on the counter you are standing in front of and proceeds to roll precipitously toward the edge. You first notice its intransigence when something slips through your fingers on the way to the counter, and many thoughts race through your mind, the first being that, given the shape of the raisin, somewhat oblong and wrinkled, the counter might not be its final destination. The evaluative undertaking you encounter might look something like the following.

There it goes rolling on the counter and, wouldn't you know it, as luck or Murphy's Law would have it, it is bounding and making a beeline toward the edge in a defiant way. And sure enough, there it goes over the edge, where gravity once again takes over—a robust natural law to be saluted—causing it to plummet toward the tiled kitchen floor. On its way, you imagine several scenarios. Once it makes contact with the floor, its rolling days will shortly come to an end and you foresee multiple possibilities. Initially, you could attempt to intercept the dried grape on its trajectory, but given your own proximity to the counter's edge, you judge this as a less than optimal solution, for you might just as easily negotiate contact with it yourself and receive blunt force trauma for your efforts.

---

2. Whitehead, *Process and Reality*, 169.

This eventuality is unappealing to you as you reckon the health of your forehead to vastly outweigh the value of the raisin. So you will need to skip to the next approach. In despair, you acquiesce to the prospect that the raisin will in fact hit the floor, leaving you with two main options. You are reassured that you need not be concerned about the well-being of either the floor or the raisin since both are resilient, provided of course that the raisin is not stepped on, for then all three of your footwear, the raisin, and floor will become casualties. Not an enticing outcome. But drop and hit the floor it must, in accord with the physical laws of the known universe. Yet, allowing for the soft structure of the raisin, at least you need not worry about the noise causing a disturbance to anyone in earshot. It is a raisin after all. Nor will it shatter into pieces, since soft bodies are above that sort of thing.

Now granted, you will need to adopt a domestic posture and clean up the unfortunate substance that hovers between a natural fruit and a candy. (I am then reminded of a personal tale involving a Homeland Security official while crossing the border into the U.S. He asked if we had any fruit with us. We responded no, only raisins. He then proceeded to claim that they were in fact a fruit. Amazing. We, however, have yet to find a single raisin in the fruit aisles of grocery stores or even in the canned fruit section. Anywhere. Ever.) And you could do this with or without a tissue, the former doubling what is already wasteful.

Though in any attempt to salvage what would otherwise be dispatched to the trash bin, you could wash off the offending agents from contact with the floor so as to resume your healthy snack; for you are convinced of the pseudo-wisdom of the "five-second rule" and are not thereby deluded. Once the raisin has hit the floor, the damage is done. The five-second time frame offers you no reassurance that the raisin is not now tainted and you would confidently opt not to simply pop the rebel raisin into your mouth without due sanitization. But then, in deference to the environment, while the raisin would thusly not be wasted, the water employed to secure a healthy confection/condiment would itself be wasteful. Neither amounts to an enviable result. Alas, both potentialities come with unwanted cost. The price is too high in either case. When it comes right down to it and all the votes are cast and the event comes to a final reckoning, you peer with disdain upon the dark violet projectile and despair that there is no worthy solution at hand and resolve to be more careful and take a smaller handful next time.

All of these things can race through your mind and they occur rapid-fire while the raisin is still in flight. We would be overwhelmed by the sheer magnitude of data impacting us if our filters were not selective. This is what Whitehead is attempting to impress upon us: the thrust of our experiences in these matters, all of which stem from the past; our synthesizing and integrating them into the current situation; our own values as applied to the circumstances; and the drawing of all the foregoing into a final conclusion, which is stored away for future reference, with the prospect that on future occasions you will know better what to do. Live and learn. The time it takes to reach a decision, whether or not the process rises to the level of active consciousness, is one event in Whiteheadian terms. Also, parenthetically, to the extent that volition has a visceral component, the above is more than simply an illustration of the (cognitive) way the brain works or education occurs.

FORCE OF HABIT

Another thinker to whom I am indebted is the cell biochemist/physiologist Rupert Sheldrake. He deals with the issue of morphology or forms of organisms and how it is that they come about and endure. As the past is important for Whitehead, in that it holds the data of the objects which have an influence on the experiencing and developing entity in the present, so too does Sheldrake's view of the past exert a shaping influence on organisms in the present. All forms had a beginning, called a morphogenetic germ, and when they did they vied for ascendancy and supremacy in typical evolutionary ways. But there is more, for ordinary natural selection was not the only factor. The very presence of a form is enabled, enhanced and carries with it a morphogenetic field, which assists in the shaping and is passed on to future generations. The more often this form appears, the stronger its field becomes and the more stable its shape.

The field acts through morphic resonance from past fields of the same form, providing as it does a type of collective memory, and through repetition assumes a stable structure. The most stable instance of this is the hydrogen atom, which has not changed its structure since its inception, nor is it likely to. Its form endures. Another example is the three-dimensional configuration of the protein molecule in some erythrocytes, in this case red blood cells, known as hemoglobin. This biomacromolecule has a specific structure together with an elaborate enfolding pathway in order for it to arrive at its final shape. Other pathways of equivalent energy could have been taken, yet they do not take place.

Sheldrake suspects that the route which hemoglobin takes has been laid down by its morphogenetic field through the morphic resonance from past forms. This resonance now rings so clearly that it does not veer from its adopted path but becomes habitual. Hence there is for Sheldrake a presence of the past, where the past acts through field and resonance so as to shape the present. He views this as analogous to a magnet, which projects a field noticeable in the way in which iron filings line up in it. As the pattern is repeated each time with new magnets, so biological structures line up in the manner in which inculcation from the past now imposes upon them in the present.

Form, however, is not the only feature so influenced. Behavior is also shaped accordingly. Like form, behavior which is adaptive gets passed on to future generations, though once again DNA is not the only factor. Genetic programs are often appealed to as a sufficient description of how forms and behaviors come about, yet Sheldrake sees an additional mechanism at work. As genes by themselves are insufficient to explain the folding pathway of hemoglobin, so genes are not fully capable of explaining certain behaviors.

How spiders, for instance, are able to spin a web without formal training is not fully contained in the genes. Nor is the migratory behavior of some butterflies and birds. As for monarch butterflies, the ones which repose in the same groves on their journey as those before them did can be generations removed from the last ones. And some birds, like the European cuckoo, "are hatched and reared by birds of other species, and never see their parents. [T]he adult cuckoos migrate to their winter habitat in Southern Africa. About a month later, the young . . . also migrate to the [same] region . . . , where they join the elders. They instinctively know . . . when to migrate . . . and . . . in which direction they should fly and where their destination is."[3]

Instinct is ordinarily pointed to by the biological community as covering all the bases, but it is not enough. This is giving a kind of power to genes, in Sheldrake's understanding, which they do not possess. Rather, once a behavior is adaptive, such as young ducklings following the first organism they see, assuming that it is their mother, it becomes reinforced such that a groove, which he refers to as a chreode, is forged in the landscape of the morphogenetic field and with repetition becomes the go-to behavior for that organism.

Once again, past behavior shapes the actions of organisms and the more often it occurs, the more habit-forming it becomes and the more stable it is likely to remain. Like Whitehead, Sheldrake's ideas are not entirely original, for they draw on previous sources, but they do give theirs a more current spin in light of contemporary science. Sheldrake for one offers several clever experiments to support his claims, some of which he expects could literally change the world. Incidentally, in some of his writings Sheldrake makes explicit reference to God as well as, approvingly, to the work of Whitehead.

---

3. Sheldrake, *New Science*, 26–27.

WHO MADE GOD?

Speaking of God, I dispensed with the art of apologetics—a reasoned defense of the faith—long ago, but I cannot resist the invitation when it comes to this question. There are those who assume that by uttering it they have put an end to all argumentation and confidently believe it is unassailable. The case, now having been closed, permits us all to retreat to our own personal enclaves and carry on with our business, untrammeled by the need to engage in needless metaphysics. The trouble, however, is that the question is simplistic, as is the sophomoric one of: if God is so powerful, then can God not create a rock so great that God cannot lift it? Presumably, this puts God in a bind, namely, which power is more ultimate—God's or the rock's? Luminaries such as Thomas Aquinas have informed us that God is not required to perform the logically impossible, like create a square circle. In the case of the rock, either there is an irresistible force or an immovable object. Take your pick. Besides, there are other models of divinity in which God's goodness trumps God's power. We can then ask the question as to which of these two attributes is the most worthy of adoration.

Another response to the manufacture of God quandary is the one about whether mathematics is invented or discovered. The numeral 3 had a beginning thanks to the Arabic world, but the number three, or its concept, reaches farther back. Much farther. Was there threeness in an abstract sense before there ever was one in a concrete form? Before there were mathematicians of any description? Before there was a world of nature? Mathematicians tend to be rationalists in siding with the faculty of reason over empiricists, who understand experience as affording the most reliable avenue to knowledge. Rationalists might argue that there is some realm from which threeness becomes instantiated into objects of observation.

Perhaps Plato was on to something. Sometimes what we allow for ourselves conceptually, like threeness, exists even prior to anyone who could imagine it and it awaits our discovery. At the same time, this courtesy is not extended to God, such that, as per Anselm, God is "a being than which nothing greater can be conceived." As threeness is beyond our capacity to "divine" its origin, so too can God be. So who made threeness? Similar problem as with God. Triangles have three sides regardless if any exist. So who made triangles? From a certain angle, pyramids appear triangular, yet notice what just happened—an appeal was made to something that existed prior to pyramids in order to make this comparison. Can something exist abstractly in some realm of Forms or Ideas before it is ever found in nature? The difference, of course, is that theists claim that God exists not merely conceptually but actually. Yet the issue concerns manufacture and no one and nothing originally made threeness, or triangularity, or God.

A different approach can also be taken. There are some who hold to a multi-universe, or multiverse, idea, that many universes exist, some perhaps even bubbling out from previous ones. In such a case, while this universe had some beginning, a universe of some description has always existed. A major problem with this position is that it is unfalsifiable—one cannot devise an experiment that would overturn the theory—meaning it is not scientific, and hence an article of faith, that is, metaphysical. For a theory to be scientific, there needs to be at least the possibility of its being refuted; if nothing is allowed to count against the claim, it becomes vacuous. Regardless of whether the mathematics points in this direction, that by itself does not constitute evidence or observation, or despite its elegance readily translate the abstract into the concrete. One might believe that in due course the experimental evidence will support the theory, but until such time, if there ever can be such a time, it remains conjectural, and scientists are not the only ones who can speculate.

There are those things which are either provable or unprovable in principle or in practice. The Higgs boson is an example of the latter. The math pointed in this direction, but there was no experiment that could reinforce it. Then came the Large Hadron Collider, where the math met the evidence, and the difficulty became solved in practice. An example of the former is string theory, where no experiment can be drawn up so as to gain evidence at such an extravagantly small scale. Despite this barrier in principle, the mathematicians are undeterred and hold on to their cherished beliefs. But this situation cannot develop beyond the math. Strong suspicion does not amount to evidence. On a larger scale, superstring theorists claim that there are branes—short for membranes—which occur in parallel, and should they come in contact at some point, a universe is then thrust into existence. Still, it seems religious-like ruminations have crept into science by the back door ("superstrings simply must exist"), despite the language and discourse allegedly being strictly scientific in nature.

Or take another level of cogitations. An infinite regress of universes could be considered an acceptable cosmological hypothesis, but is one more example of a view unprovable in principle. We do not even have direct access to the entirety of our own universe, but only the visible part of it, a small proportion of the whole. We find ourselves in a light cone which indicates that very section of the universe where light signals can reach us. Most of it cannot. Only light can carry these messages. No signals, no knowledge.

Much of the remainder of the universe could be antimatter for all we know. Though there are some assumptions that we can make, such as the temperature of the universe, called the microwave background radiation, which, after all, as part of the electromagnetic spectrum of radiation, is a form of light. We can assume this to be in a state of equilibrium—the same everywhere on average, known as isotropic—if the thermodynamics to which we are accustomed is universal. If so, we can speak about the (at least to us) invisible universe in this way even though we have no access to it.

Now please do not misconstrue what we are advancing. We are not using science, it must be stressed, as a way of proving God's existence, for that would be a form of natural theology, which suffers from its own weaknesses. Arguments which end with the conclusion "Therefore God exists" ultimately do not work, and regrettably the history of Christian scholarship is riddled with them, even today. The main issue is that the conclusion does not point to any particular divinity. Which deity is represented among the candidates of Yahweh, Allah, or even Brahman is sadly inconclusive. What I am insisting on, rather, is that what is allowed for science, by way of hypothesis formation, should not be at the same time disallowed for religion. If what comes under the umbrella of science can be speculative, the same ought not to be ruled out for religion. And since science at times is conjectural and tentative, religion ought to be extended the same courtesy.

As we have seen, science can have religious contours and religion can have scientific. An example of the latter is allowing archaeological findings to inform the reliability of one's religious beliefs. If one has faith in the regularity of nature and therefore science, one should procedurally or methodologically allow the same for religion. Faith is not a bad thing, for everyone exercises it, from scientist to religionist. This does not permit us to make any specific theological claims we wish, but it does confer the opportunity to engage in the discipline scientifically. This makes religion a little bit more scientific and science a little bit more religious; and we ought to be accorded the opportunity to conduct religion scientifically and science religiously, without fear of censure on either end. And science should be unfettered and disabused from the danger of adopting the posture of a superiority complex.

Back to our main issue. Inconsistently, the question we began with is regarded as a knock-down argument. That universes might always have existed is a laudable claim; that God has always existed is assessed as fanciful, even farcical. Once again, some individuals allow for themselves what they do not allow for others. That this can be extended to the moral realm is a topic for another discussion. But at least permit me to say the following.

Natural selection, to introduce another concept, deals with survival, but not the truth of one's beliefs. There is no guarantee that our cherished and dearly held inclinations are true, for from an evolutionary psychological perspective they assisted in the survival of our ancestors. And this success rate is all that matters to evolution. There is no biological test for truth, only survival and the passing on of more genes in the form of offspring to the next generation than one's competitors. Thoughts and ideas are either adaptive or they are not. The difficulty is patently plain in the fact that both theists and atheists have survived. Both have a long history. Even the Psalms proclaim that there are atheists about: "The fool says in his heart that there is no God" (Psalm 14:1; 53:1).

Admittedly, the time span between the psalmist and now is insufficient for evolution to occur, but I am assuming that then was not the first time that anyone awakened to the prospect of there being no divinity in heaven or on Earth. Yet both theism and atheism cannot be true. So for the simple, bald impetus of survival, either one may get you there. This tells us something about their survival value and nothing about their veracity. Perhaps it does not even tell us this much, and maybe the proclamations of evolutionary psychology are overextended. And without an actual subject upon which to experiment, our distant ancestors are all gone and are not saying much, except for the artifacts and fossil remains they leave behind. These can tell us something about what they did and how they behaved, but not much about what they believed. To be buried with weapons or food can indicate what the deceased expected they might require in preparation for the hereafter, but it also might simply inform us as to what their favorite keepsakes were. ("He loved that dagger, so let's bury him with it.")

Thus far I have not come across a compelling argument as to why atheism should prevail over theism. Perhaps the exercise to do away with God is fueled by the attempt to rationalize one's own personal preference and even establish one's own autonomy. The same cannot really be said, though, about the theistic option. In Christianity, for instance, followers are called to a life of discipleship which may involve sacrifice and/or suffering. Aside from those with masochistic tendencies, it is usually not one's natural inclination to invite them.

NOW THAT I THINK OF IT

No one knows when self-awareness or self-consciousness put in an appearance. Perhaps we will never know. The least we can say is that the onset of these did not leave any artifacts; there were artifacts before and after, but none which definitively mark the origin of a reflective being. Yet there are comments that can be made about it, and some thinkers have done so in scientific and philosophical history. Alfred Russell Wallace, a contemporary of Charles Darwin, was one of these. The mood at the time was such that more and more events previously thought to have their cause in the outworking of the divine will were then reckoned to suggest a natural explanation, to the point where little if anything could remain outside the realm of natural forces. Far from arguing which few events were not the product of divine activity, the burden of proof came to lay at the feet of those claiming that there were in fact still some that are. Nature, not the deity, was seen to be the cause of what appears.

Wallace, though, thought otherwise. He viewed the human mind not as the result of natural selection, but as a direct divine deposit. At one point hominids had a brain but no mind (for those who make a distinction between the two) and then at a subsequent point, thanks to God, they bore both. Wallace could not fathom that the capacity to reflect on the eternal verities was generated through natural forces alone; there must be a mind-maker behind it all.

Most thinkers did not follow him in this, though there are still some around even to this day. The majority, however, suspected that nature itself was the culprit. If Wallace's diagnosis is not the correct one, then which is? The knee-jerk response of "There can be no explanation but a natural one" is insufficient. Even if true, this does not put an end to the work which needs to be done. One must ask, for instance, how self-awareness/consciousness fared once it surfaced. The idea that once it emerged it also thrived and was, like rock and roll, here to stay might be wide of the mark, or at least hasty.

Consider a hominid bearing the trait, perhaps alone amongst her peers, noticing herself as different from them and desiring that they should join her in her newly discovered attributes. Recognizing that her efforts were futile, she departed in an existential fit of rage for parts unknown and ultimately perished without offspring. An evolutionary dead end. Or perhaps, again, there were two such creatures in the same vicinity which, upon evaluating the merits of boiling versus barbecuing, came to blows over their differences and met a similar fate.

The point is, once self-awareness/consciousness made its initial foray into the world, its success was not guaranteed. It boasted faculties which others did not possess, but this did not automatically make it adaptive. "So what if you can, as you claim, ruminate about the beyond? Cogitate about this!" The new capacities might not have been welcomed by others, or maybe even by those who first self-identified with them. "What should I do with this? Oh, what shall I call it—angst?" Others might not have seen a use for it or considered it an improvement and thus it may not have flourished. "There goes Gronk again with his head in the clouds" (although it could have required higher cerebral powers already so as to make such an assessment).

It could be that self-awareness/consciousness needed multiple appearances before it became successful, that is, led to the leaving of more offspring with the same traits. It would be overreaching to declare that its superiority to life without it was readily apparent. A case would need to be made that the better way ahead lay with it than without it. In hindsight, we may surmise that its advantages are obvious, but it might not have started out this way.

If Wallace's view is inaccurate, then we are left to wonder not only how it got its start but how it blossomed. And even if he were to have been correct, God's work was also not done. The divine will may not have been met with a single injection but multiple attempts. Or, alternatively, if the event occurred but once, then it might have needed constant divine nursing and attention, ensuring that the spark was fanned into flame amidst all the potential perils of existence in nature's "rouge in enamel and talon." Either way, self-awareness/consciousness had its work cut out for it. Was the experiment, whether natural or divine or both, worth it? At least now we get to reflect on it.

As an aside, there is a debate about whether we are a) a divine product with the aid of the evolutionary process or b) an evolutionary product with a desired divine direction. In essence, does God's general plan rule with evolution in charge of the minutiae, or is the evolutionary program dominant with God filling in the details? Of course, evangelicals have the position of divine sovereignty to defend, so their response will be: "God has authoritarian rule over everything, for there is nothing outside the divine purview." Liberals have no such stance to uphold, so they are freer to consider the alternatives more openly. And the possibilities hinge on the extent to which

God not only enjoys sovereignty but elects to exercise it. If God relinquishes power (the openness strategy), then God can allow free creatures like ourselves to exercise some of our own as well as allow natural selection some latitude; but if God does not wield a coercive type or amount of power at all (the process platform), then God's prominence is shifted to the background, leaving option (b) as the lone possibility.

This brings me to a possible scenario.

HARRY (HAIRY) THE TROGLODYTE

In the beginning there was a creature upon the Earth, a forerunner in our own ancestry. The sapiential character of this organism was without form and pretty well-nigh empty, about as pronounced as a voided check. Yet its spirit (Hebrew *nephesh* = breath) ran deep, for the breath of God animated its body, and it hovered over wherever it could find fresh water and it brooded over its family circle. It was driven by God's wind (Hebrew *ruach*) to survive. The darkness of a cave was its home, for therein lay safety.

But the creature thought, on behalf of its family unit, "Let's go into the light, for I hunger and thirst," knowing full well the risk of exposure to the elements as well as to predators ready to pounce. "I expect that we are vulnerable on every flank," it bemoaned, if it only had the words to express it. There was hunting and there was gathering, another day. And the creature (let's call him Harry, for he was still hairy) thought, "I am deeply troubled, for the fires that we make to cook our food fills our cave domicile with a not inconsiderable amount of smoke, so let's take our act outside. I trust that animals will not approach us where there is a fire." They attempted it, it worked, and it was good. The end of a successful day.

Come morning, Harry thought, "I bet if I had a tool kit, I would increase my catch and decrease my work load (since we are all about conservation these days)." So he scraped one stone with another until, behold, it took an edge, and fastened it to a long stick and made a spear, although he did not have the wherewithal to name it that, or anything else for that matter, for ease of reference, since he was still at the prelingual stage, so he realized he would need to resort to charades should it ever go missing. Filled with hope, he ventured out to secure victuals, since there was not as yet a corner store stocked with comestibles, but found that critters would not hold still long enough so as to be pierced. He despaired that "we are perilously close to just vegetarian fare tonight." So he opined, "Enough of this," and took a stone in a sling and slew a hare at long last.

That was acceptable, but, exasperated, Harry hit upon the idea, "Maybe we will be more successful if our community consolidates its efforts and each able-bodied person fashions a spear and surrounds our prey, with the prospect, of course, that each will share in a fair and proportionate distribution of the consumable goods." So Harry and sons (much later to become a landscaping crew) went off with their hunter compatriots and captured a larger animal. But then they realized, "Ya know, advantages also come with their attendant disadvantages, for if we drag it back to camp, we will attract some uninvited guests simply by the aroma of the carcass." Hence they decided to dismember the beast into more manageable portions to haul back to camp in a less burdensome manner and leave the remains for scavengers and sundry decomposers. Complex reasoning for the time, when you think of it. Upon arrival, they were greeted by the other family members, who wanted to know if their exploits were favorable. Harry concluded, "It was good (can't you see?)." And Harry surveyed all that he had—family, food, and fortress, in that order, although the alliteration was lost on him—and he reckoned, "This is very good; real good in fact."

Later, Harry looked around and surmised, out of unenlightened self-interest, that there could be more to life. Dissatisfied with all that he had, he sought greater contentment—not a place in the "burbs," mind you, for there were none as yet to be had, no urban-rural distinction or gentrification or any social stratification for that matter. Yet observing that the sandy soil appeared to be browner for the neighbors down in the valley, he wanted more and was willing to fight to get it, thinking that they would be unwilling to relinquish it voluntarily. This he did. He and his sons turned their spears into weapons of warfare and killed another of their kind as they would any other prey.

They made off with the booty but were struck with a feeling of unease. "We got what we wanted," but they experienced a constant need to wash their hands, in time their booty lost its lustre, and they diagnosed that it was like something inside had been torn. They could not escape a sense of existential fear that something irreparable had been done. It wasn't hunger, for food could not fix it. Was this a type of early-onset PTSD? (They didn't really think in those terms; that's just an embellishment.)

The rest of human history is predictable. Ideas lead to inventions, advanced technology leads to more innovation, and power leads to corruptibility. If there is a way to distort and pervert something, somebody will find it. But the story has its upsides too. There are remarkable acts of selflessness and charity; beauty abounds in visual art and music; and the world itself astounds with the diversity of life. We live in an extravagant place. Harry and family were in a position to recognize and benefit from some of that as well. We are part of what we have come to appreciate. Though with comforts arise anxieties, and life and living is full of the tensions between them.

SCIENTIFIC ADDENDA

As an addendum, in a previous volume I indicated that despite our best efforts at curbing the effects of natural selection, with advancements in medical science for example, they are but Band-Aid solutions, for natural selection in winnowing out the weak and preserving the strong will win out in the end. But this is only half the story. Now for a theological perspective. The description just related is emblematic of a distorted creation, which God will restore. God's care for the poor and needy is a direct counteroffensive to the effects of natural selection as well as the political systems that would seek to support it.

Nature by itself—and we are part of it—is in the business of reinforcing Machiavellian policies where might makes right, and those who can survive the day deserve to perpetuate their lineage. But God is in the business of heeding the cause of the disenfranchised and downtrodden, together with ultimately settling accounts with those who facilitate this action. If natural selection can afford to be patient, since it will win out in the end, God will redress the elements causing these grievances, making them yield to God's transformative pursuits. Ultimately, God is the one who will achieve the final victory over extinction, for natural selection is actually not the final word. When God draws history to a close, survival will ostensibly be God's domain. What precisely it is that survives is another issue to be addressed in due course.

On the topic of natural selection and God, take the following gastronomic example. Humans delight in good food. The joy of a fine meal is unparalleled by many other enjoyments, but good taste is not always good nutrition. Salty, sweet, and fatty tastes tickle the pallet, but that is where the benefit ends. The drive tends to be toward increasing the intake, perhaps even to the point of uncontrollable urges, and yes of course I will totally have fries with that. We seem to have developed a taste for the unhealthy in our long evolutionary history. We have become geared to stocking up on these tastes since they were often in short supply. The difficulty surfaces when we do not know when to stop. Whenever we greedily ingest those foods with such gusto, the results become apparent. The drive might be to survive, but it has gone too far. We have been built to preserve energy, but have overdone it to our own detriment. Natural selection prepared us for survival, but now with such foods readily accessible, we do not always employ the off button.

There are centers in our brain for both appetite and satiety, but the latter is easier to ignore and suppress. From a theological perspective, God has bestowed upon us culinary pleasures but sometimes our god is our stomach and we descend into gluttony. Something that God intends for good becomes distorted such that the focus of our attention is not directed toward properly managing our diet. As to who or what is to blame, where does the responsibility lie? Is God to blame for granting us something that can be so easily perverted, or is natural selection the culprit for not as yet catching up to what currently constitutes survival? Obesity can have both genetic and cultural components; the question is to what extent there is a volitional aspect to this and whether such behaviors stem from the choices we make. Are we victims or offenders? From where does the urge to reach for the chips ultimately arise? In the meantime, while I ponder the matter, please pass the gravy.

As another addendum, I hesitate to buy in to the late Stephen Jay Gould's program in his work *Full House*. His notion is that organismic excellence or extremes of abilities in evolution are diminishing and in its place an average is being approached. As time goes on, variations in abilities decrease and plateau. While this may be true for some capacities, others as a whole show little sign of succumbing to this condition. He employs, among other things, a baseball statistic in order to reinforce this claim, namely batting averages. Specifically, Ted Williams in 1941 was the last player to hit over 0.400. Gould envisions this as never to be equaled, let alone eclipsed. While granted in this case, other statistics could be used which do not support his thesis so readily. Records continue to fall in the Olympics, both the summer and winter games, revealing that extremes are actually being approached, not departed from.

But staying with baseball, the velocity with which pitchers have been hurling has increased over the years, where the 100-mph barrier has periodically been surpassed. Why not use such a statistic instead? Well, because it does not help your cause or bolster your assertion. Also, in football the distances from which field-goal kickers have been able to split the uprights have been increasing steadily, overcoming as some have the sixty-yard mark and beyond. Why should batting averages constitute the main sports index? If one doubts this movement toward excellence, one needs only to consider the skills competitions in the all-star games of the various team sports and observe how often previous records have been supplanted. Moral of the story: test theories to determine if they can withstand scrutiny, for the metric utilized may tell but one side of the account. Hence I urge the reader to, as a proper journalist would, investigate all sides of a story.

Finally, genes are sometimes understood as omni-explanatory and able to account for any part of us. But here are some things DNA cannot explain: it offers no rubric as to why I am a sports fan or keep tabs on the teams I do; nor why I was a dualist (where reality is made up of two things: body and mind/soul/spirit), became a psychosomatic unity proponent, and then back to a modified dualist (a theme already mentioned and to which we shall return); nor why I have the favorite fiction and non-fiction authors that I do. It has been claimed that there is no accounting for some people's tastes. To an extent that is true, genes notwithstanding.

RELIGION-R-US

The knock on religion in postmodernist quarters is often directed towards the violence perpetrated by metanarratives. The overarching account of a religion's history betrays a trajectory of violence to some group or other. The Christian metanarrative, as feminist deconstructionists have informed us, does violence to women, for instance. Religion has been roundly castigated for its violent metanarratives and we should thereby seek to avoid them whenever possible, they would contend. Some feminists think this situation can be remedied and therefore the religious baby need not be thrown out with the violent bathwater, while others are not so optimistic. The postmodernists have carried this conversation for multiple generations, yet is the avoidance that they insist on really feasible?

It has recently come to light that the makeup of humans is to engage in narratives, to weave together the pieces of our history into a seamless cloth. Alasdair MacIntyre has alerted us to this tendency on our part, where we are authors who have inserted ourselves, our personhood, as the main character into an arc or account and have thereby become "story-telling animals." Others have countered that what we are left with are what may be termed "memory quanta" and we should leave it at that without attempting to connect the dots into a meaningful whole.

In essence, there is no continuous person, self, or soul; we are just in the habit of concocting one. David Hume and Buddhists, for example, would applaud this view and it has both religious and legal ramifications. Religiously, we are not in a definitive position to declare that there is an essential component to us which survives death and participates in an afterlife; and legally, if a person is not continuous and hence not the same from one moment to the next, then are we warranted in prosecuting a later and perhaps then unrelated, or less-than-fully-related, version of someone?

But there is more. Current neurological studies have enlightened us that this is how the brain works: the unconscious sifts through a massive amount of quantized material and the conscious artfully molds a sample of it into one piece. Consciousness has evolved to seek continuity where perhaps there is none. This must have had a selective advantage at least in the past and maybe even today. The foregoing allows us to go beyond postmodernist insights and side with the penchant of which MacIntyre speaks and posit the following: biologically speaking, we craft narratives and cannot (at least as yet) do otherwise. Strenuous conscious effort might not resolve this. And those who are disgruntled about it should blame natural selection.

These steps appear to fall into place: to be human is to engage in narratives; these sometimes become crafted into metanarratives; and these in turn have a religious character about them that often do violence to some group. As a musician plays a piece of music, the move has been made from notes on a page to a melody, given our habit of erecting a structure from its parts. Analogously, our consciousness makes the move from individual episodes to a fully formed person, self, or soul, which are philosophical-religious categories. This now has scientific corroboration. The point to be made is that it does not do any philosophical good to ask humans to refrain from being religious in some sense, for our current biology will not permit it. Correspondingly, the upshot is that philosophy-religion will need, as a reminder, to take science into account.

23 AND WHO?

(Disclaimer: Please note, the following advertisement is fictional and bears no intentional resemblance to any person, living or dead. A fuller explication will follow.)

> We at ethnicity.com celebrate the fluidity of cultures. Those who avail themselves of our services bask in the knowledge that an accurate investigation into their cultural background may reveal membership in an ethnic group not previously recognized and thus free themselves from the stereotypes that have so shackled them up to this point. Clients refer to these encounters with their new natures as conversion experiences. One Irishman exclaimed that upon the realization that he was actually more Icelandic than Irish he was thus freed from his bellicose nature and traded in his boxing gloves for skates. He also became more gentle and easygoing, though sadly he ran out of luck.
>
> Another man, a Sicilian, discovered that he was mostly Danish and thus felt more contented with his life and no longer experienced the need to hold a grudge. Best of all, a Muslim man uncovered that he was more Jewish than he realized and turned in his prayer mat for a shawl and no longer felt the need to campaign for Palestinian rights.
>
> No doubt about it, cultures dictate who we are and we cannot evade them. Just ask both of our satisfied customers (the Icelandic fellow got too cold).

The above is largely bunk. The television commercial in which a formerly suspected German man unearths that he is more Scottish and opts for a sartorial change from Lederhosen to a kilt is still the same person and likely does not automatically switch cultural mores upon such an awareness moment. Nor does the Scottish man who discovers that he is more Italian suddenly decline his savory haggis (sheep's bladder) and develop a taste for Mediterranean cuisine. One does not exchange one set of stereotypes for another so readily. If genes completely dictated our tastes, we would not be so quick to make the above changes if we are initially convinced otherwise. More on this presently.

IDENTITY AND ME

The question of what constitutes the essence of a person together with what precisely it is that crosses over from one life to the next raises a number of additional issues. For instance, is it essential for, say, an ethnically Jewish person to be Jewish? Plato aside, no decision was made on the part of the individual, either while in the current life or prior to it, to be cast into the role of a member of a certain ethnic group. Other categories are at least partly chosen: social, cultural, political, religious, and philosophical affiliations and even behavior can alter over the course of a lifetime, but even if a Jewish person were to convert to Catholicism, s/he would not cease to be Jewish ethnically.

Personally, I am a first-generation Canadian of German immigrant parents, and no matter how many times my parents were to have emigrated, or where to, I would not cease to be of German heritage. This is one of my lots in life. The question is whether this belongs to my essence. In genetics, one's genotype is the set of chromosomes one bears and the genes (together with the alleles) they contain; the phenotype is the extent to which they are expressed in one's physicality. But who a person is seems to run deeper than even genotype. So which part of us can be Jewish or German? Being either of these, it seems, is not completely accounted for on the basis of genotype. Or phenotypically, having blonde hair and blue eyes does not make one German; wearing Lederhosen even less so, for these are outward manifestations. Take me out of my context and I remain German, but to what extent?

To belong to a certain ethnic group is not to have an affiliation unless one participates in it culturally. For me to be me, I cannot avoid some German particulars—it is part of my makeup (or duty?). Yet this does not by itself press upon me to undertake the stereotypical activities of drinking beer or dining on knackwurst. Thus genetic ancestry services err when they advertise that knowledge of one's forebears catapults one into a change of wardrobe and cuisine. Moreover, when we speak of something that might cross over from one life to the next, Jewishness is not one of those things, entailing theologically for the Christian world that the resurrected Jesus is no longer a card-carrying Jew, for otherwise, in classical Trinitarian terms, one of them would not be like the other two. Being German, or anything else for that matter, might be a part of who we currently are, but our essence lies deeper still.

Perhaps taking another tack will help us. Curiously, some patients who have undergone organ transplants have later reported that they have experienced a change in likes and/or dislikes. In the after picture, a person might possess a new inclination for having his or her chicken Kentucky fried. Upon further investigation, it may be disclosed that the donor of the organ bore this penchant. Mystery solved, though it presents us with further issues requiring explanation, for something from the essence of the donor, in this case an inveterate taste which does not switch gears so readily with changes in custom, can at least partially be contained in a body part and crosses from one person to the next.

Hence it might be warranted to consider choices made in this life to figure in to the essence of a person, and in turn implying three things: first, Jean-Paul Sartre may have been right all along, at least in this regard; second, in line with his thinking, our essence reflects becoming rather than being, phraseology which process exponents would applaud; and third, this time a question, if our essence reflects our choices, then can we expect to have similar inclinations in whatever shape the next life will take?

To elaborate, Sartre sees us as the sum total of our choices and our essence is constructed thereby. On the one extreme, Plato takes essences as residing in the realm of Forms, which are then instantiated in organisms that are about to come into existence and are a pale representation of them. The eternal natures are the Forms themselves; the transitory natures are the beings they imperfectly reflect. Hence with Plato, essence precedes existence. On the other extreme, Sartre understands existence as occurring first; we find ourselves in the world whereupon we must choose, and the choosing produces an essence. Consequently with Sartre, existence precedes essence.

Whitehead, on the contrary, emphasizes becoming over being, and whereas Plato's essences are in the realm of Forms, Whitehead's existences are all in the past and essences are embedded in those objects as well as in the mind of God. Becoming is in an unsettled present; being in a settled past. Essence for Sartre is the set of decisions made; for Whitehead it is the set of experiences encountered and responded to. And note that neither Sartre nor Whitehead even so much as mentions genes as contributing to our essence.

To take the discussion one step further, creativity would not be something that belongs to our essence but instead happens to us. Inspiration occurs to us; hence it does not reside in us. It does not originate with us and therefore would not cross lifetimes. If one finds oneself in the Platonic camp, then creativity would be a Form that could be channeled to an individual;

if one is more in the Christian tradition, then it would be something that God's Spirit could bestow. In either case, we cannot lay claim to it; the responsibility for it lies elsewhere.

On the theme of crossing lifetimes, consider the following. There are reports of reincarnation where the old life brings with it to the new a striking similarity of birthmarks, which are external features, in addition to internal aspects such as personal predilections. If both reincarnation and those accounts of it are accurate, then room will also need to be made for outward manifestations to cross over. The stigmata of Jesus' resurrected body as attested to in John's Gospel (20:27) would seem to concur with this. His risen body could have been free from defects, yet not according to this account (or are they badges of honor?).

Now neither reincarnation nor resurrection theoretically precludes the other. And whereas both could be in effect over the course of a lifetime and beyond, there does appear to be a sequence of before and after. Resurrection could follow reincarnation, but not the reverse. Resurrection is the likely end point, though this would not automatically rule out becoming. If the work of John Hick is true, who champions the insight of the church father Irenaeus, we can expect two processes at work, one succeeding the other. In this life we are fashioned in God's image, which includes the capacity for self-reflection, relatability to God, and participation in God's moral attributes; in the next we will be further transformed into God's likeness (2 Corinthians 3:18). The work of refinement of essences, then, would not end with death.

PRELIMINARY CONCLUSIONS

It is an inevitability of the creative process and perhaps, in tune with the theme of my previous volume on human nature, to become dissatisfied with one's finished product after an initial sense of accomplishment sets in. No more has this occurred for me than in the foregoing work, owing largely to what I omitted. I can shrug off a few peccadillos, but I cannot ignore the critical mass of omissions from the previous study, prompting what is for me new territory, namely, an afterthought. Whereas a preface allows me to say what I really want to say, a postscript enables me to reflect, after the fact, on realizations that have come to light in the process. So, breaking with the convention that insists on a postscript only at the end of a text, here goes.

To begin with, as the apostle Paul claimed in 1 Corinthians 15:51, "we shall all be changed." But changed into what? As was intimated, we will trade perishable, corruptible, mortal bodies for imperishable, incorruptible, and immortal ones. We inquired as to the nature of these new bodies and whether there is, or the extent to which there is, a continuity with our old ones. Compounded with this emerges the debate about whether not only our exterior but interior qualities, namely, our character and memories, also make the trip. Those who favor resurrection differ from those who side with reincarnation. Memories, it will be recalled, cross over for the resurrection camp but not for the reincarnation group, at least from a Platonic viewpoint. This means that the selves that we are in the "after" portrait are different from the "before" picture, depending upon which category we find ourselves in. A self with an intact memory is different from one that lacks it.

The point is this—and it becomes an issue only for the resurrection crew, for only they are committed to the following implication: if the transformation we are to undergo in the resurrection event is so radical that we can only describe the before picture of the body as physical and mortal and the after picture of the body as spiritual and immortal, then to what extent can we still be understood as properly human? We will be more "like the angels" (Matthew 22:30; Mark 12:25; Luke 20:36)—creatures with spiritual bodies—who are not human. But the fun does not stop there, for this leads to yet another, this time even more alarming, ramification.

Strikingly, if our former situation is human and the latter is not, then God's intention from the beginning was toward the shedding of our humanity. Reformed Christians have argued that a Platonic understanding of Christianity must be wrong since dualism devalues the body. Recall we found that a dualistic posture need not take on a Platonic dimension. Yet if this version of resurrection is accurate, then the human body is in fact downgraded as the dualists claim, for it eventually outlives its usefulness. More precisely, dualists contend that bodies are prison houses in this life, a stance which is not supported by Christian sacred texts. In our reformulation of resurrection, then, human bodies are valuable, but not ultimately so. Evidently bodies are not so bad even from a spiritual perspective, for note the preoccupation on the part of unclean spirits to seek to inhabit them (Matthew 12:43–45; Luke 11:24–26).

Hence we need to be clear about what it means to be human and how much change we can withstand before we lose it, for it has a staggering significance beyond this life. Does humanness cease at death? Is it merely a stage that we pass through to an eventual not just different but higher state of existence, since the spiritual trumps and maybe even beats the daylights out of the physical? It must, because that's where we are allegedly headed. Otherwise God intends for us a lesser or lower final end. That's not progress, and that's not like God.

I stated in the text above that I learned something about myself, namely, that I am more dualistic in philosophical and theological orientation than I realized. The foregoing seems to provide additional fodder for it. Our bodies are expendable and disposable; so too, it appears, is our humanity, at least as we know it. Apparently we are made to lose our humanity, as the shedding of a snake's skin or, more accurately, the metamorphosis from caterpillar to chrysalis to butterfly, where the before is significant and good in its own right, but cannot match the after in beauty, and so is discarded.

I find it unsettling to initially argue, as I have in the past, for the value of humanity, only to discover that it will later give way to a new and improved type of existence. I fear that the Platonic dualists were partially correct all along. I am compelled to agree with them that we can anticipate a higher state in the life to come. Where we differ is in the non-Platonic view that humanness does not amount to our being incarcerated, but is important while it lasts.

A second point I wish to raise here involves another aspect of human nature as revealed in the Judeo-Christian Scriptures. In Matthew 11:18–19 (and Luke 7:33–34), Jesus bemoans the reception which he and his relative, John the Baptizer, have received: "For John came neither eating nor drinking, and they say, 'He has a demon.' The Son of Man came eating and drinking, and they say, 'Here is a glutton and a drunkard.'" These two situations, on both extremes, appear to exhaust the possibilities—either people eat a variety of foods and drink a variety of beverages with gusto or they do not. In either case, there will be critics.

No matter what you do, in certain circumstances, you cannot win. If someone or a group has decided—and it very much amounts to a decision—that s/he or they do not like a given person, then they will view whatever the individual does in negative terms. They will be faultfinders, even if there is no fault to find. They have simply erected a barrier to acceptability and have chosen to be implacable, turning the Pauline encouragement on its head: "Be discontented, whatever the circumstances" (Philippians 4:11). This is the way we operate sometimes, damaged goods that we are.

Third, I purposely avoided a certain topic in my discussion about the mind, but I can forego it no longer. Experiments have demonstrated that when persons are asked to make particular choices in a test setting, they believe that they act freely without any encumbrance, yet specific brain sites are activated, revealing that the choice has already been made, even seconds beforehand. The subconscious has dictated how the conscious will perform, all the while permitting the latter to imagine that it has acted on its own accord without any prompting. As Freud might exult, "See, the subconscious really is in charge and determines our selections and actions." The line between science and philosophy has thereby been crossed in stating the seemingly unavoidable implication: "So there is no free will after all." This conclusion, however, might be hasty.

One must first take a step back and examine how the subconscious has arrived at its own decision. If it can be traced back far enough, and as Rupert Sheldrake has informed us, there may have been a time when such a neural pathway was not yet established, but came into formation by the initial selection on the part of the bearer. An action may have been judged appropriate by the actor and been reinforced by repetition. By virtue of its inculcation, it then becomes habit-forming so that a future decision need not be made, but can be relegated to automatic mode. A person might still believe that a decision is being reached afresh, even though an acceptable choice for him or her has been made long ago. This might be how we have survived, by committing to reflex or instinct what needs to occur so that we need not deliberate about it each time. The choice is then predictable for having initially been free.

Fourth, I regret to inform my readers of the following scientific issue, and it is not about the inevitability of our developing some form of cancer if we live long enough. Rather, my pessimism about human nature is matched by the cynicism surrounding a dissolving universe. Here are the details.

Ninety percent of the fuel required to form new stars has already been spent. This means that the fertile ground of the movement from simple to complex as described by the evolutionary process has reached its peak and we are well into a downward trend. This decline is marked by the paucity in new star formation, which supplies us with the raw materials framed in the periodic table of elements and manufactured in the interiors of such large stars. Fewer stars, fewer resources.

We are facing a shortage of the goods necessary to form complexity, meaning that all we have in front of us is dissolution. The universe has already experienced its springtime blossoming and summer fullness; harvest time is nearing completion and all that remains is a late autumn chill and the plunge into the deep freeze of winter. This is not to say, though, that the universe is old and "On Golden Pond." The universe is still quite young; it is only suffering from the cosmologically equivalent condition of flesh-eating disease. It is becoming decrepit, and the winter is never to be replaced by springtime once again.

The cyclical universe of Eastern religious traditions is not borne out by astrophysics, since we are in for a frozen eternity. This will not occur for an exceedingly lengthy period of time, but unless the cosmos will behave in a totally new and unexpected way, or there is an injection of workable energy from outside (outside where?), then the cosmology is inescapable—the world and everything in it is destined to become a cold, dead corpse (from which no one will be around to pry away the late Charlton Heston's firearm).

Whatever we have made of and with our human natures can encounter no end in this scenario other than a footnote in cosmic history. Our epitaph will read, "Ambiguously harmless and harmful, benevolent and malevolent." But there will be no one to cast a gaze upon it. We are doomed. (Unless, of course, God has an emotional investment in the world and seeks to renew and transform it.)

Depressed yet? First time diagnosing cosmic physiology? A little disappointed, a little perplexed? This is the bad news, yet the good news is that the end is a long way off, so there is still time to take that trip to Maui. The record of our salutary as well as deleterious effects on the planet is still being written. Hopefully the beneficial will outweigh the injurious; and perhaps we can make our little plot of ground into a garden and enjoy it for a season, while there is still time, for time will tell.

Fifth and finally, my hope is that the afterlife will be akin to the aftermath of a football game, where the players from both sides gather to greet and congratulate each other on a match well played, extend well-wishes, and give each other updates on their lives. The heat of the battle is gone; there is no more competition now that the final whistle has sounded and this particular game is in the books. I trust that, just like in the sporting event, if there was any resentment in this life we can move beyond it and even share a chuckle about it when all the smoke has cleared and dust settled. From this vantage point, when all historical events are simply that—in the past, there is nothing left to contest, so we can get on with the business of communal relations, where the most common refrain will hopefully be, "No problem." I just wonder whether we will be any better at social interaction than we are currently and, if so, how this would come about.

I'm Herb Gruning and I approve this postscript.

PART 2

Laying the Groundwork

TEXT AND THE NEW CREATION

Now that we have tested the waters, for terrains can also contain them (think of the Everglades or Minnesota), the time has come to develop earlier themes. Allow me to begin with a warm-up exercise. Recall our foray into the classical divinity's metaphysical attributes. In a previous volume I discuss how they fare in the light of current thinking. Spoiler alert: not well. But this is not what we are about here.

The moral attributes will serve as a springboard for our purposes. God's moral qualities portray the kind of being God is believed to be, and, unlike God's metaphysical properties, we both can and are called to participate in them, hopefully in ever-increasing measure. They are named by eleven or so character descriptors. In alphabetical order they are: faithfulness, goodness, grace, holiness, justice, love, mercy, patience, righteousness, truthfulness, and wisdom. In contrast to the aforementioned metaphysical attributes, the moral ones require no elaboration or elucidation since they conform to the standard dictionary definitions. Or do they? When we use these terms, do they mean the same things for us as they do for the deity? Is there a one-to-one correspondence between human and divine expressions of them? This is to ask whether they are univocal—one voice for us as well as for God?

There is a partial answer for this. At the very least, there is a difference in degree between our reflection of them and God's. God, for instance, is much more patient than we could ever be. But what needs to be asked is whether God's expression of it is purely more of the same sort that we are familiar with, or is it also of a different type? To be different in degree is a quantitative distinction; to be different in kind is a qualitative one. Which is it when it comes to God's character?

C. S. Lewis informed us that the English language is impoverished in regard to the number of terms devoted to love, for it has but one. The Greek, by comparison, has about four, one of which describes God's type of love, namely *agape*—a self-sacrificial love that expects nothing in return. This is not the same as our ordinary understanding of love, since it is the love of God, thus another term was needed to convey its meaning. It is qualitatively different from ours, though God wishes us to develop to the point where we can emulate it as well.

This exercise of differences in degree versus kind is pertinent for one of the tasks that lie before us, specifically the nature of the new creation and our place in it. No doubt there will be differences both in degree as well as kind from our customary scene. The Apostle Paul in 1 Corinthians 15 informs us that there will be both continuity and discontinuity from our era to the next. There will be the familiar of sorts in addition to the extraordinary. Paul employs the analogy of a plant and its seed to communicate the transformation. A physical seed is required to produce a physical plant, although plants look nothing like the seed. The genotype is the same, but its phenotypic expression is starkly different. There is both a continuity of substance and a discontinuity of form from one to the next.

Is it reasonable then to assume that there will be some familiar counterpart to the standard natural laws in the next era that we have grown accustomed to and operate with in this? Is there anything we can say even by way of speculation, knowing full well that our limitations of language and imagination prevent us from obtaining anything resembling or amounting to full disclosure? Are we prevented from catching a glimpse, even if the reflection is poor and our sight veiled, given that now "we see through a mirror dimly, but then face to face" (1 Corinthians 13:12)?

But there is something else to this exercise. God's attributes may or may not find reinforcement in the Judeo-Christian Scriptures, yet it is this very Bible that we must examine as to its usefulness as a resource. Does it help or hinder our quest? Another spoiler alert: it is ambiguous—sometimes helpful, sometimes not, for it contains contradictory accounts. I outlined some of these in my prior two volumes, but the reports were not, nor were they intended to be, exhaustive. There is more to tell. We need to consult these inconsistencies so as to evaluate the type of text we are dealing with.

APPROACHES TO SCRIPTURE

There is a certain brand of Christianity known as evangelicalism. By way of introduction, evangelical is not the same as evangelistic, which refers to the program of attempting to share the gospel message in the hopes of converting their hearers into the fold. Rather, the evangelical anchor to the faith is the Bible, which they laud as the authority and final arbiter of faith and practice. For them, the Bible is not one resource tool for the faith, but *the* resource. Some background is called for.

During the waning years of the nineteenth century and the early years of the twentieth, there was a group of Bible believers, beginning with some in the Niagara region, disgruntled by recent developments in the sciences. Among other advancements, which they did not view as such, were the fields of geology informing us that we live on an old Earth and biology telling us that we arrived here not by having been specially created but through the evolutionary process. In both cases, we are the product of natural forces and causes. Appeal thereby need no longer be made to received authority or metaphysical explanation. These developments troubled the evangelical crowd, whose commitment was to the Scriptures and the direct-from-God messages they contain.

Their response to this crisis was to draw up a manifesto as well as to establish Bible schools and colleges as a defensive strategy against these "doctrines of devils." They composed what they believed to be the Five Fundamentals of the Faith and a tract of the same title. From this they drew their name as fundamentalists, originally a descriptive as opposed to a pejorative term. What topped the list of these five essentials was their perspective on the biblical statements: they were to be understood as infallible and inerrant. They contained no mistakes on any theme they addressed, including creation, taken as a six-day divine event. Their shibboleth to identify those who belonged to this threatened enclave was not only the biblical message in general but particularly the first eleven chapters of the first book of the Bible, Genesis, which contained pertinent information on the origin of the Earth as well as humans (and every other creature for that matter) in addition to the Noachic deluge (the flood) and the tower of Babel. No other source need be consulted; the Bible is sufficient.

It might surprise them to learn, however, that not even they believe their own rhetoric. Despite playing the literalist card whenever opportune, they too recognize that the Jesus of John's gospel, for instance, is literally none of a lamb, a shepherd, a vine, a door, or a narrow gate. Hence even literalists know when they come across metaphor, which by its very nature does not lend itself to literalist interpretation. Nor should rigid theological conclusions derive from, for example, poetic passages, as the same can be said for this methodology as well. To make such a move is not without its drawbacks, for by admitting that parts of the Bible can be taken literally and others not, conservatives are unwittingly perpetuating precisely that which they reject in liberal strategies.

Now there are multiple positions one can take on the Bible and they can be placed on an axis or spectrum. On the right extreme, chosen because it is the conservative stance, resides this fundamentalist posture, where whatever the Bible says, God says. The Bible is entirely a divine product. On the other extreme is not this literal camp but the liberal one, where the Scriptures are to be treated as any other sacred text and even any document from the ancient world, like Plato's *Republic*. Works from the great world religions contain valuable insights from an ancient cosmology, replete with myths that science can now answer, so no special pleading is allowed. On this end, the Bible is purely a human product.

In the middle could be called the inspired position, for there the Bible contains statements ignited by the religious experiences of humankind, complete with the gifts, talents, skills, and abilities of the authors, together with their five Fs: fragilities, frailties, flaws, foibles, and failures, in increasing order of negativity. In this middle position, the Bible is seen as a combined effort—both a human and divine product. Evangelicals appear on the right half of this continuum and fundamentalists on the right extreme, meaning that evangelicalism is the broader category. The authority given to Scripture is greater the further right one goes. Evangelicals need not adhere to the literal view of Scripture, but they can. If they elect to do so, then they take their place at the right extreme. Hence one can be evangelical without being fundamentalist, but not the reverse. Fundamentalism is a sub-camp within the larger camp of evangelicalism.

Humans can be found at all points on this axis. A cross section of society lies throughout the ranks: some people with mortgages, some who take their kids to soccer practice, in fact, the same type of people you would find at any other point. Yet that which defines them in terms of this continuum is the authority they grant to their sacred code of belief and conduct. When engaged in theological discussion for those on the right, the task becomes one of unlocking what the biblical position is on the matter at issue, for there can ultimately be but one such posture, and once it is uncovered that becomes the end of the debate, though each denomination will have its own interpretation. For anyone to assert that this might not be what the Bible teaches or that it could also be wide of the mark is to be cast adrift as non-evangelical at best and an unbeliever or heretic at worst, with all which this entails.

I for one do consider the Bible to contain ambiguity at certain places and error in others, and in so doing I betray my own position. These are more than just the quirky and idiosyncratic, and they range from the trivial to the profound. An example of the former is found in John 9:4, which announces that the "night is coming when no one can work." In the ancient Near East, nighttime prevents work from being carried out, but this no longer holds from the late nineteenth century onwards; just ask someone on night shift. Electricity enables around-the-clock activity; casinos bank on it. Despite the potential metaphorical intent of the passage and important matters of context, it became obsolete ever since the Chicago World's Fair, known as the Columbian Exposition, of 1892–93, where the wonder of streetlights illuminating the night was bestowed upon us. Yet such an instance of obsolescence is not integral to the thrust of our argument; it just further sets the stage.

DISCREPANCIES AND INCONSISTENCIES

In a previous volume, I outlined biblical statements which affirmed the invisibility of God, opposite others which confirmed just the contrary—an instance of profound ambiguity. To amplify on this theme for a moment, 1 John 4:12 flatly proclaims that "No one has ever seen God," while in the Old Testament, Numbers 12:8, partially along with Deuteronomy 34:10, maintains that "With [Moses God] speak[s] face to face . . . and he beholds the form of the LORD" (NRSV), and in the New, Hebrews 8:5 presents the same: Moses "persevered because he saw him who is invisible." The Bible makes it difficult to adopt a position when it argues both sides of an issue.

Another example was the perseverance of the saints: the notion that "once saved, always saved" protects one from falling away after having initially been redeemed. Once again, an ambiguity. The same amount of passages can be produced on either side of the debate, thereby settling nothing; the problem remains. Denominations take their stance on this doctrinal question by emphasizing only half of the statements and ignoring the other half, or explaining them away. The result is that they do not believe the entire Bible, nor is it even possible to do so. The ability of humans to rationalize the agenda they feel they need to defend is truly astounding.

An additional volley into the theater of the erroneous is glaring, since it contains historical inaccuracy from an author who chronicles events and is not unaccustomed "to set[ting] down an orderly account" (Luke 1:1–4). Should Luke be the author of both the gospel attributed to him as well as the second volume—the book of Acts, which covers the events of the nascent church in addition to the travels of Paul—then he should have taken care to be more precise, especially if he is so interested in truth, as he outlines in his prologue.

We are informed in Luke 2:1–2 that a registration was decreed by Emperor Augustus Caesar at the time of Quirinius. Jesus was born near the end of the reign of Herod the Great, who died in 4 BCE, but Quirinius became governor of Syria in 6 CE, ten years later. Luke has the expectant couple, Joseph and Mary, journey to Bethlehem, at which time she gives birth to the child. Yet the timing is off, for Jesus could not have been born at that point since he already would have been about ten years of age. Should the translation as stated in most versions of the Bible be correct, then either Luke is mistaken, or, what seems more likely, the rendition is intentional and he has a purpose for doing so.

Quite possibly, he wishes to portray the events surrounding Jesus' birth as being consistent with Old Testament prophetic accounts, in this case Micah 5:2, where it is held, though not definitively, that the Messiah is to arise from Bethlehem. Luke was keen to market Jesus as a prophet and as such he would need to be demonstrated as in accord with honored prophetic writings, so he needs to get him to Bethlehem realistically somehow. The trouble is, among other things, that there likely never was such a census as he describes. It is much more probable that Jesus had his birth, childhood, adolescence, and adulthood up to the age of about thirty, when he began his ministry (Luke 3:23), in Nazareth in Galilee, a region very different and distant from Bethlehem, unless of course one holds to the notion that he traveled to exotic lands on a religious quest in his formative years. Yet even here, Luke might be betraying his agenda in one more way by having Jesus appear as in the prophetic or at least major figure tradition, in line with both Joseph (a son of Jacob), who became governor of Egypt under Pharaoh at age thirty (Genesis 41:46; 42:6), and David, who became king at age thirty (2 Samuel 5:4).

One of the most famous discrepancies is one that helped fuel the Protestant Reformation. Luther came to the realization that "the righteous will live by faith" (Habakkuk 2:4; Romans 1:17; Galatians 3:11; Hebrews 10:38) and that "we have been justified through faith" (Romans 5:1). Luther took as his ammunition the many references which Paul makes to these declarations in his letters, particularly, but not exclusively, in Romans. It must be pointed out, however, that nowhere in the Scriptures is the term "alone" included; nevertheless, Luther inserted this sentiment as though it were intended by default. The Reformers in general supported the ideas of faith alone and the Scriptures alone (*sola fidei* and *sola scriptura*, respectively), as a type of back-to-the-Bible movement. Champions of this tradition they took to include Augustine, meaning the Reformational emphasis occurred well before later evangelicals got hold of it.

The debate between faith and works heated up at this time. The fear on the part of the Reformers was that the more one stresses works, the more it is made at the expense of faith, such that one could actually merit reconciliation with God. They need not have been so anxious, for other Scripture passages were not recommending works over and above faith but alongside it— faith plus works. The epistle of James promotes the notion that "faith without works is useless/dead," reasoning that deeds complete faith (James

2:14–26). Reinforcing this viewpoint, the gospel of Matthew endorses the injunction to "prove your repentance by your deeds" (3:8; also Acts 26:20b), and John the Baptizer is held to have fostered the same: "Produce fruit in keeping with repentance" (Luke 3:8), implying that once one has faith one should put it to work. Paul exclaims that we have received reconciliation precisely "in order that we might bear fruit to God" (Romans 7:4b), and even goes so far as to alert us that "the only thing that counts is faith expressing itself through lov[ing deeds]" (Galatians 5:6b).

In essence, deeds can be seen as a necessary but not sufficient condition for being reconciled to God. Thus we can look upon works as a gift given to God for God's own gift of life to us. They become our grateful, thankful, freewill offering to God as a pleasing, aromatic sacrifice, since we do not want to go to a party, in this case the wedding banquet of the (Passover) Lamb (God's Messiah), empty-handed (Exodus 23:15), for that would not be good form.

But wait, there is more. Consider yet another instance, this time from the Old Testament wisdom book of Ecclesiastes, where the teacher is convinced that persons "can do nothing better than to eat and drink and find satisfaction in their work" (2:24), together with Isaiah 30:15, which states that "In repentance and rest is your salvation." These two passages do not speak about loving works of faith and service whatsoever. Hence not only are the Scriptures ambiguous at this point but they hold at least three distinct positions (amtriguous?) and theologians are left to deliberate over a priority list. And if they disagree amongst themselves, which they do, quite vigorously and even heatedly at times, then where does that leave the rest of us?

A further example us the belief of the early church that Jesus' return was to be imminent. Noticing that this had not yet occurred, it becomes an understanding of "soon" of which we are unaccustomed to hold. When generations passed, there was a re-evaluation of the doctrine and followers interpreted it as intending to say that soon in the deity's terms is not the same as ours, or that the divinity was giving people an extended opportunity to turn to God. This overlooks the fact that the text can also be interpreted as meaning Jesus' current generation. There are those both within and without Christendom who chide the Jehovah's Witnesses movement for their, to date, eight failures to accurately decipher and predict history's culmination: 1914, 1916, 1920, 1938, 1942, 1961, 1966, and 1984 (some of these years and the events they contain being admittedly ominous), but is even one instance on the part of the early church to be excused? Perhaps one can say that they were new at this. Most importantly, can a mistaken view actually be evaluated as inspired in the first place?

Either way, the upshot of all this is that one will need to look elsewhere if one wishes to find a set of documents free from error, provided there is such an elsewhere to speak of, for the sacred text known as the Judeo-Christian Scriptures will not provide it. They contain a wealth of spiritual resources of the kind we would all do well to heed, but not the sort of accuracy consistent with modern standards. The authors certainly were not aiming for this, nor are they required to, for that is a recent innovation, but then they fall short of the very accuracy they uphold.

Some miscellaneous comments: The high priest whom the author in Mark 2:26 has in mind, namely Abiathar, is not the one mentioned in the passage to which he refers, for in 1 Samuel 21:1–6 the high priest in question is Ahimelech. Not a careful reading or recollection. Plus, in the two episodes of Jesus feeding the multitudes—five thousand males in addition to women and children in Matthew 14:15–21; Mark 6:35–44; and Luke 9:12–17, and later four thousand in Matthew 15:32–38 and Mark 8:1–9—there were twelve and seven basketsful of leftovers collected, respectively. Questions worth asking include: Who would be following Jesus in this somewhat remote region with baskets in hand? Just being prepared as a Boy Scout would be? And what would have been their contents? Were they empty? As another example, since Jesus was in the habit of healing the crowds that followed him, why then when he went into the home of Simon the leper in Matthew 26:6 and Mark 14:3 did Simon apparently remain a leper after the visit? There is no indication that his condition had improved. Or was his name merely a moniker? Still another, the three Synoptic Gospels agree that Simon of Cyrene was forced to carry the cross of Jesus (Matthew 27:32; Mark 15:21; Luke 23:26), while the Jesus of John's gospel carries his own (John 19:17). So which is it: did he or didn't he? Or was it just partway?

Next, there is a discrepancy between Acts 15:37–40, where Paul is disinclined to take Mark along in his itinerant ministry because he had previously deserted them, and the Paul in 2 Timothy 4:11, who urges Timothy to "Get Mark and bring him with you, because he is helpful to me in my ministry." Now Paul, of course, could have had a change of heart, or only one of these versions at most could reflect his actual mentality. If the former, he likely would have made mention of his erstwhile displeasure together with his about-face. The final option is that Paul is not the author of the Timothy epistle and the sentiment was stated on his behalf.

In contrast, I do find that several incidental and some parenthetical details in addition to certain off-hand remarks can be relied upon, such as the following. Exodus 9:31–32 gives instructions as to when four different grain crops ripen; Isaiah 28:23–29 submits fun facts on how to farm crops (to be fair, this is for the purpose of alerting the people that this wisdom comes from God); Exodus 16:36 offers information about weights and measures; Deuteronomy 3:9 informs us as to how different nations refer to Mount Hermon; and 2 Samuel 11:1 makes the claim that spring is the time of year when kings go out to war. We are given no reason to doubt these statements, since they neither support nor detract from the writer's agenda (with the possible exception of the second of these five), though, as we demonstrated above, there are other names and historical details about which we cannot have the same degree of confidence. In the New Testament, Mark 5:41 informs the reader of the meaning of an Aramaic phrase in Greek, permitting us to surmise that the audience of the gospel is in fact Greek and not Hebrew. Reinforcing this supposition, Mark in 7:3-4 offers an extended discussion of Jewish purity observances, leading us to affirm that his audience was not familiar with them.

Completing our scrutiny, the resurrected Jesus of Luke 24:37-9 invites his hearers to "Touch me and see; a ghost does not have flesh and bones, as you see I have," which is at odds with 1 Corinthians 15:50, where Paul declares that "flesh and blood cannot inherit the kingdom of God." Plus, while we are mentioning Paul, a Jew, he angrily exclaims in protest against the Jews that "From now on I will go to the Gentiles" (Acts 18:6), only to promptly go "into the synagogue [to reason] with the Jews" (verse 19) and "[enter] the synagogue [to speak] boldly there for three months" (19:8) in the subsequent passages.

It is entirely possible that these and other changes mark a development in the reasoning of later authors from previous ones, but the point remains that there is a difference which reveals an inconsistency regardless of the rationality of its onset. In essence, the Bible is a useful tool which can be mined for its nuggets of wisdom and it affords insights into what certain authors thought about Jesus and God. But it needs to make up its mind on certain topics, please.

FOOLISHNESS TO THE MULTITUDES

Both Psalms 14:1 and 53:1 inform us that "Fools say in their hearts, 'There is no God.'" Evidently there were fools, that is, atheists, then as now. A current understanding is such that belief in God and the Scriptures was taken for granted in the ancient and medieval worlds and perhaps only the meaning of the text was at issue. In the history of Christendom, church councils convened to discuss and hammer out what should constitute orthodoxy in theological problems. Yet it appears that not everyone always took everything for granted. Skepticism lives and has lived.

It would seem to be more accurate to claim that the Bible exemplifies a "minority report." It was actually the fruit of a collective conservative mindset and many if not most were not on board. It could very well have been written by a priesthood interested in the status quo and who wanted to retain their advantageous position. The writer(s) of John's gospel declare that the statements contained therein are offered in order "that you may come to believe that Jesus is the Messiah" (20:31a). The masses decidedly did not, otherwise such an appeal need not have been made. At least he was up front about his motive.

Were we to ask if something is historical, we would intend whether it can be evaluated as factual. Biblical scholars urge that this was not a preoccupation on the part of the early communities. They told stories which packed a punch, the details of which changed sometimes so as to suit the circumstances, all in an attempt to build up the evangelistic weapons in the arsenal of the storytellers. The intent was not to render the facts and only the facts, for that is a contemporary strategy. Rather, the pursuit was to frame the message in such a way that it would have the greatest impact on its hearers, in the tradition of Jesus as a master storyteller and provoker of discussion and debate.

To question one's belief system is not impious but part of a healthy faith. Yet is the insistence on the part of some that others need to be argued into the fold in a philosophically sophisticated way a sign of health or a symptom of the psychologically unwell? Participants in apologetics—a reasoned defense of the faith geared to winning over its hearers—undoubtedly feel they are filling a need, meaning they must be spurred on when facing those they deem "foolish" and the alleged misconceptions they hold. Or are these in the target audience simply seekers and honest questioners?

I agree that it is inappropriate to impose one's beliefs on others if it is in the spirit of believing one has "found" that which others do not know they need, for it is futile to give answers to those who are not asking questions. This applies to both parties. What is the point for those assumed to have found the truth and think they no longer need to ask questions to attempt to give answers to those not asking them? To make matters worse, no one invited them to initiate the discussion. There are churches and helplines to assist with this. Other than that, keep it to yourselves unless you have earned the right to be heard.

There are a variety of reasons why atheists are around. Our job is to ensure that theistic believers do not provide these atheists with those reasons. The right for anyone to believe what they wish must be respected. Until of course they ask questions, for then we have the opportunity to search for answers together. Best to adopt a perspective of mutual searching; this gives both parties ownership of the quest. Apologetics engages in providing justification for one's belief; though not all recipients of this endeavor have an appetite for it and would rather avoid the diatribe. An outbreak of sobriety might suggest that doubt can be productive; imposition not so much.

Perhaps an acronym for the more apt program should be S.M.A.R.T. (Society for the Mutual Advancement of Reasonable Thinking), as opposed to S.P.I.T. (Society for the Prevention of Independent Thinking).

THE JESUS OF THE TEXT

There are different ways to portray Jesus and every gospel author writes from a different vantage point and has a different agenda concerning this exemplary figure. At times the accounts are similar but at others they vary markedly, so all we have is the Jesus of any given text, not the actual historical personage. The four gospels contained in the New Testament—and there are many others that did not make the cut—were written approximately forty to seventy years after the events they cite. During this period, eyewitnesses drop from the scene and oral traditions become modified to suit the circumstances and audiences at the time. The upshot of all this is that the exercise to conclusively establish not only what Jesus said but did and was like is not straightforward.

The initial step to take for our purposes here seems to be opening up the text as it stands. Even before deciding whether the accounts are reliable or what amount of confidence can be placed in them, we need to be clear as to how they read. Given that contemporary biblical scholars ordinarily are not also theologians and vice versa, the latter normally work with the writings themselves, side-stepping for the most part issues of literary criticism in the volumes they produce. This can be an oversight, but not everyone comes with multiple areas of expertise. As it turns out, theological categories might not be so safe after a little scrutiny of the text.

Let's take the example of the doctrine of omniscience—the extent of God's knowledge— and focus on how this is portrayed in the life and ministry of Jesus. If he is to be understood as the second person of the Trinity, then the incarnation left him with only the moral attributes of God and not the metaphysical, because if he were truly like us then he could not have possessed these latter properties, since we don't either. If he were God in disguise, having full access to the perquisites attending to divinity despite being human, then he was not one of us, but an alien. These appear to be the only two alternatives, and there seems to be a slight ambiguity when consulting the text.

On the one hand, Jesus is said to know the thoughts of others. Mark 2:8; Matthew 9:4; and Luke 5:22, known as parallel passages, together with Matthew 12:25, state that Jesus knew in his spirit what others were thinking in their hearts. We are not informed as to whether he had this ability due to an alleged divine status, or if he knew it telepathically, or if he was intuitively aware, or if he was just a good guesser. On the other hand, there are many more passages which indicate the opposite, for he poses many questions. He asks who touched his clothes (Mark 5:30b; Luke 8:45); how many loaves the disciples had on hand (Mark 6:38; 8:5; Matthew 15:34); he inquires as to the length of time a certain individual had been suffering from an ailment (Mark 9:21); what his followers were arguing about (9:16, 33); and most importantly admits that no one, not even he, knows the time of the end except for the Father (Mark 13:32; Matthew 24:36).

Aside from the last of these, Jesus could possibly have played the part of an attorney in a court setting asking questions he already knew the answer to, but in common parlance this would eventually become tedious if not also outright annoying. Another item of note is the startling claim that "Although he was a son, he *learned* obedience from what he suffered" (Hebrews 5:8, italics mine). This passage implies that not only was Jesus not omniscient, he was also on a learning curve. Evidently he needed to become educated in the ways of God and that this would involve suffering. It appears odd from a classical viewpoint at least to propose that a divinity is required to learn.

Nevertheless, as it stands, the scales tilt in favor of his not being omniscient as God is, otherwise he would not need to ask these questions or make such statements. Hence the theological insistence, wherever it might occur, that he be omniscient is not so patently plain from the text, but rather seems to be foisted upon him. Perhaps definitively, the Jesus of John's gospel is emphatic about an alleged Trinity (a term which, of course, never arises in the Scriptures, but might be another example of this type of foisting) as not displaying equality when he flatly declares that God the Father is greater than he is (John 14:28). After all, there is only one particular camp within the tradition that holds to his omniscience and they do not speak for all of Christendom, although I imagine some of them would eagerly wish to. Regardless, none of the above needs to detract from his being Messiah.

Another item concerns Jesus' character and, keeping in mind our attention is directed at the text, whether he displayed a short fuse. This time God's moral attributes are in view and a temper is not what one would expect if Jesus also had them. That would further place in jeopardy the standard depiction of Jesus as gentle, meek, and mild, essentially a wimp. Here are some of the major indices.

The text paints him as talking back to his parents as a twelve-year-old (Luke 2:49–52) (despite the following statement that he was, or became, obedient to them and grew in favor with others); he ignores his mother's and sibs' entreaties to speak to him (Mark 3:31–35; Matthew 12:46–50; Luke 8:19–21) (they believing that "he is out of his mind" [Mark 3:21]); he resorts to name-calling, specifically "hypocrites," aimed at the scribes and Pharisees (Mark 7:6; Matthew 15:7), and "fox," in reference to Herod (Luke 13:32); and he becomes exasperated with his obtuse disciples (Mark 9:19; Matthew 17:17; Luke 9:41). We are also warned against calling someone a fool (seemingly out of anger) (Matthew 5:22), yet apparently there were exemptions for God (Luke 12:20), Jesus (Luke 11:40; 24:25), and Paul (Galatians 3:1), all of whom employed the term. Another inconsistency.

And though the next two side-by-side episodes have a significant prophetic intent in that they prefigure the destruction of Jerusalem (Luke 19:41–44), they nevertheless convey personal characteristics: he curses the fig tree, even though the creation, particularly the ground, already stands under God's curse (Genesis 3:17–19) (thereby making this a double curse?), since it bore no fruit for him—an unreasonable request since it was not the season for figs (Mark 11:12–14; Matthew 21:18–19); and he clears/cleanses the temple by driving out vendors and overturning their tables and chairs (Mark 11:15–17; Matthew 21:12–13; Luke 19:45–46). John's gospel is even more specific by adding the detail that Jesus used a whip made out of cords to motivate them (2:15). Perhaps all together these passages indicate that he was subject to mood swings.

There are those who would examine his ministry and diagnose him as irritable. (Some claim that we become like the god we believe in.) To make matters worse, Jesus is further depicted as not above using deception (John 7:8–10). As it turns out, this is a caricature, for Jesus is also portrayed as loving and compassionate. The negative characterization is partially offset by passages such as Matthew 11:29, where Jesus states that he is gentle and humble in heart, and 12:18–21, claiming that "a bruised reed he will not break." The latter case, it must be pointed out, is taken from Isaiah 42:1–4 in reference to God's servant, who is the nation of Israel. Only later did the author of Matthew make the connection that this could also apply to Jesus. Additionally, when Jesus enters Jerusalem, he does so on a donkey (or its colt) and is described as gentle (NIV) or humble (NRSV) in Matthew 21:5, though once again this is taken by Matthew as a fulfillment of prophecy, this time from Zechariah 9:9. But the point to be made is that he is not without his anger management moments. After all, we can seldom be described as being one thing ourselves, like cartoon characters, but are ambiguous as well, alternately positive and negative. Hence the idea that we are given a sanitized version of Jesus in the gospels does not always hold.

Having said all of this, it remains that Jesus not only had a thorough knowledge of the Old Testament despite being uneducated—"How did this man get such learning without having studied?" (John 7:15)— but radically turned accepted, conventional wisdom—what everybody already knows and takes for granted and as such is never questioned or examined—on its head. He employs the vehicle of parable to this end, as a means to draw his listeners into the story and come out on the other end realizing that they were in fact the part of the story that needs attention. In this way, Jesus was a master teacher and storyteller, for if you do not want the negative aspects of your character exposed, then it would be best not to listen. That is why Jesus said, "He who has ears, let him hear" (Matthew 11:15).

On a further personal note, I wonder if the human Jesus had any regrets in life. Sure he lamented, even wept, over Jerusalem, which killed the prophets sent to her (Matthew 23:37; Luke 13:34; 19:41); of course he bemoaned being misunderstood by family members and others (Mark 3:21); and most assuredly he had trepidations about his imminent mistreatment at the hands of the Jews and Romans. Anybody would (even though Luke's gospel paints a picture of his being rather unflappable about it all, despite the passage highlighting his having sweated something "like drops of blood" while praying on the Mount of Olives [Luke 22:43–44], likely being a later addition by an editor seeking to mitigate Jesus' stoicism and making him appear more truly human, emphasizing his role as a prophet as opposed to a divinity). But did he also have reservations about how he conducted himself? Did he ever wish he would have said or done something differently and did these memories haunt him? Peripherally, did it ever hurt him to hear, if and when approaching a new town with his disciples, "Here comes the God squad," due to his reputation preceding him?

Paul and Barnabas, for example, "had such a sharp disagreement" about the value of Mark to their ministry that they decided to break up the team and go their separate ways, with Paul taking Silas along and Barnabas associating with Mark (Acts 15:37–40). Perhaps some things were said in the heat of the moment before their departure which they could not take back and this became a source of grief for them. Did Jesus have similar unguarded moments or incidents, whether as a child, adolescent, or adult? If so, I suppose there would be contemporary counselors who would like to know how he worked through them. Yet the overarching question might be whether Jesus would have recognized himself in the portraits painted of him.

It needs to be kept in mind that the writings come from a time period well after the recorded events had taken place and accordingly they reflect the era in which they were drafted. What this means is that, for instance, those Jesus had in his crosshairs, namely the Pharisees, were not as perceived at the time of Jesus, since they did not wield such power until after the fall of Jerusalem, almost forty years after Jesus' death, and were granted their authority by the Romans. The Pharisees at Jesus' time did not enjoy that degree of power, whereas those at the time of the gospel writing did. This was part of their agenda, to place the Pharisees in a negative light, for the community had been persecuted by being thrown out of the synagogue for their allegiance to Jesus, and this did not occur until toward the end of the first century.

Another crucial matter is the fact that Scripture is not given; rather it is made by a devoted community. We decide what constitutes Scripture, and these writings contain agenda-driven material. Take for example the passage in Matthew 27:51–53, where at the point of Jesus' death some graves of the faithful were opened, some bodies were raised and appeared to many people. In essence, there was a resurrection prior to Jesus' own. Something this monumentally important, one would urge, could not be overlooked as insignificant so as not to be included in the other gospel accounts. Yet they are not there! The same occurs with the raising of Lazarus in John 11:43–44. How could these episodes fail to make the editorial cut or escape the notice of others? The reason might very well be that they are imaginary and served the purposes which the writers of the gospels intended.

The issue is not only that the biblical authors wrote from their own perspectives and agendas for their own audiences, but that we do the same things when we enter the fray and open the text for ourselves. It is not just a matter of placing strenuous effort into approaching the writings objectively, for this is an ideal beyond the capacity of anyone who admits that s/he possesses subjectivity. Pure objectivity does not exist and it is simply not humanly possible to keep oneself out of the calculations. We can commit ourselves to becoming aware of our own prejudices and seek to clean them up, yet even this commendable effort does not guarantee success, for we can never extricate ourselves from an encounter with a text. We and our baggage always come along on the trip. Besides, how could we ever know if and when we have arrived at seeing more clearly? How could this assessment be made? Would we not just trade one set of biases for another? We still own our subjectivity and can no more dismiss it than we can the shadow we cast outside on a sunny day. This necessity need not trouble us; we just need to recognize its effect on us and attempt to check ourselves when it occurs, all the while realizing that the task is not foolproof. The trick is to live comfortably with the tension.

PET PEEVE

Having said all of the above, and in order to demonstrate that I am an equal opportunity critic, there is also something about the Islamic sacred texts I have trouble with. Here is the source of my disquiet. In an otherwise commendable religious posture, there appear to be two sets of morality in the Muslim world; specifically, what is denied to men in this world, namely wine, women, and song (well, not so much the latter), is openly made accessible in the next. But what is it that makes something immoral in this life but moral in the one to come?

There is a difference between withholding a good thing until such time as one has reached sufficient maturity so as to be responsible with it, such as the operation of a motor vehicle, and withholding something because it ought not to be made available at all. Ever. In the second instance, issues do not cease to be but remain good or bad, as the case may be, both now and in the hereafter. In this way, there would be no discontinuity or divine flip-flop. Something does not become good merely because one has passed from one form of existence to another.

Recall that God asks us to emulate God's own moral attributes. Period. A case could be made that what is good for God may not be good for us, harkening back to the age-old question, is something good because God says it is or does God evaluate something as good because it is good? In the second instance there is a standard to which God must also conform, and in the former God is above any such standard and gets to proclaim what is good. But even, say, cruelty? In the grand cosmic scheme of things, we are never above the law, despite the antics of certain politicians, so it becomes a matter of how stringently we should uphold it.

If the law or some form of it is a permanently binding moral code to be adhered to in any form of existence, then good and bad remain ongoing categories. On the contrary, if the law can be superseded at some point, as some would argue that the life of Jesus has accomplished, or is no longer in force in the hereafter, then this still does not undermine the ability to recognize what was at least once a good or bad act. Something bad does not automatically become okay when one puts on a resurrection body. There will not be a softening or relaxation of that which constitutes right or wrong once one has "graduated" to the afterlife. As the creation myth states, the gods possess the knowledge of good and evil and desire to spare humans from that burden (Genesis 3:22, my paraphrase).

If the aforementioned behavior should be transcended in the now, it should also be transcended then, not given as some type of reward for avoiding it here. Moreover, this would be contrary to the teaching of the Jesus of the text, a prophet whom the Islamic world also honors, that there will be no marrying or giving in marriage in the hereafter (Matthew 22:30; Mark 12:25; Luke 20:35). Perhaps this is because there might not be gender as such in the world to come, a topic to which we shall return.

ENTER THE DIVINE

Now on to the crucial matter of how does or can God work in human lives in ways that make a real difference to us? In terms of a top-down approach, and invoking the plasticity of the human brain, God could "have a hand" in the shaping of the contours of our behavior. Employing Sheldrakean language, if our decision-making pathways have formed sufficiently deep grooves such that the patterns of our behavior have become habits, then God could be involved in the establishment of these patterns and their ongoing maintenance. Changes in behavior would then amount to a reorganization of certain dendrites and the neuronal pathways they represent. More on this momentarily.

The germ for such a material realignment could very well be the impression which God's spirit makes in the life of an individual, and there Whiteheadian language also applies. Should it have a marked impact, then impression becomes translated into expression. In this way, God's persistent presence can have effects, beginning at the psychological/emotional/visceral level with changes of heart and resulting in the material alteration of the brain. These modifications in turn become instantiated at the behavioral level. Here matter and psyche work in conjunction.

As for the bottom-up approach, explanation is still sought not only for the psychological but also for the physical levels. It is alleged by some in the theistic fold that God performs interventions in the world. Admittedly, many of these occur at the psychological level, with God not manipulating the materials of nature with divine tools in the divine tool kit as such, but transforming lives with changes of mind. In essence fashioning a new person: "creating in us a new heart and renewing a right spirit within us" (Psalm 51:10).

Take for example the quantum and molecular levels. The "mechanics" of God's quantum activity can be described using the familiar language of indeterminacy and wave function collapse. In some cases, microcosmic causes can have macrocosmic effects. In others not. Where it occurs, a quantum event can change a hydrogen bond in a DNA molecule from one nucleotide base to another and alter the course of evolutionary history. This travels up the scale from electron to molecule to cell to tissue to organ to organ system to organism, and in the process the expression becomes increasingly visible to our scrutinizing senses. A deity could likely concentrate on instances of either top-down or bottom-up approaches and might not be limited to one or the other.

Hence God's influence could take place in either direction, for both would be open to the divine consideration. Yet this would not account for all of the effects that are alleged to have a divine, or at least an "extra-anthropic," source. For those that remain, they await further theoretical outworking. Perhaps divine activity takes the shape of working in tandem with the body's own capacities, which it is ordinarily assumed by those in the God-friendly camp that God has invested our physical makeups with such proclivity and receptivity to begin with. God could then enhance what was present from the outset or, alternatively, restore what has become diminished, all in conjunction with already-existing conventional bodily abilities and functions. In this way, God becomes less of a foreign agent than we might initially have supposed.

On the Saturday of Easter weekend some years ago I listened to a radio program which discusses scientific issues of the day, and on this occasion it aired an interview with Brother Guy of the Vatican Observatory in Arizona. The thrust of his message was the close connection between religion and science, much more so than is deemed to be the case in the modern world, no thanks to the media, which looks for a controversial story. Brother Guy ably demonstrated that, historically speaking, at least in the West, science grew out of religion, from the desire to know the order of the world which God crafted. Guy is a creditable champion of what he holds as the congruence of religion and science.

Despite the admirable and deserving accolades he earns, I find that I have some misgivings with the content of the message he delivered. There are three main ones. The first is the extent to which God acts in the world. Guy maintains that God could act miraculously but elects not to. He takes as support for his view the biblical episode of Abraham and Isaac. The initial thing to mention by way of commentary is that it does not do any good to discuss the extent of God's activity here if the event is unhistorical in the first place, for God, if God acts at all, acts in history and not in legend. Whatever we can draw from the account, it is not a smoking gun, but perhaps a moral of the type akin to Aesop's fables. The same would be true for those biblical passages not yet confirmed through archaeological evidence, such as the Noachic deluge, the exodus from Egypt, in fact, everything prior to the earliest attested biblical tale, namely, the house of David, as written on a stele discovered in 1993. Important life stories they are, but imaginative tales nevertheless.

Back to Guy's argument. The voice which the likely fictional Abraham heard is a cause that has a physical effect. If it were to have been a voice which anyone present could have heard, then it was a physical cause creating longitudinal transverse waves which impact an eardrum. If it was more than simply "a still small voice" (1 Kings 19:12) in the KJV, or "a sound of sheer silence" in the NRSV, or "a gentle whisper" in the NIV, essentially inaudible to anyone but the intended target, a strategy which God could certainly employ, then this would constitute the work of an agent from outside the world entering it and interacting with it. In either case, neurophysiology would be affected. Neurons would fire and signals would be carried and this amounts to a physical change, which in turn seems to point to a divinity who freely participates in the world that belongs to God.

Objections using the language of this making the deity out to be yet another link in the causal chain of events appear to sidestep the issue that free will/choice and activity are traceable, as the Prussian philosopher Immanuel Kant would claim, to reasons, not causes. Agents have reasons for the bulk of their actions, despite the insistence on the part of some psychological schools of thought, like the Viennese psychoanalyst Sigmund Freud, in which the impetus of free actions are in reality disguised causes. I did not need to write that last line, but I did; and you did not need to read it, but you did. Since I was under no compulsion to pen these words and could have done otherwise, we do operate with some amount of free will. Both you and I have reasons for doing so, and they then became causes which in turn produce effects.

I agree with Guy that God acts in the world; my contention, however, is that God does so with physical effects if not also outright causes. God consistently operates in accordance with the order of the world that God has set up, not in an attempt to be consciously science friendly but so as to be efficacious. If God did not do so, the effort would be entirely futile, for it would miss us completely. The difference is that these causes would indeed have effects, though not necessarily God's intended ones, for, as Whitehead insists, they depend on the responses of free creatures. According to him, and contrary to the classical theistic view, we have the power to thwart God's intentions.

Second, Guy correctly states that humans through God's guidance have disclosed physical means in order to deal with physical problems, intending by this to declare that we must do more than simply pray that God will change that which in the ordinary course of events, if left to itself and under its own steam, will not effect the intended change. God has given us the cranial capacity and wherewithal to investigate the world and apply what is "unearthed" toward the alleviating of the world's difficulties.

Two comments. The first is why God would not have imparted some of this knowledge or have us stumble upon it sooner so as to promote the common good. Vaccines would be a case in point. We can be grateful that they were discovered at all, but this is cold comfort to those who did not make it under the wire. The second is the mentality that physical problems must have physical solutions, or that all we need to do about a difficulty is throw more technology at it. It could very well be the system itself that needs repair. One question might be not how to extract more from the Earth and more inexpensively, but how to find alternative sources of assistance and, ultimately, stop damaging the planet. The latter in particular entails an ethical reorientation.

Finally, there may be some concern as to what Guy considers to be authentic science. He welcomes scientific input and allows religion to guide his science as well as the reverse, yet is there a limit to his inclusivity? Would he foster, say, parapsychology as bearing some insights into the world or is it by definition an illegitimate pursuit? Much as universities conservatively uphold the status quo, I suspect he might distance himself from it as something "too great, marvelous or wonderful for me" (Psalm 131:1; 139:6), though the Bible offers little indication as to what this something might be, other than God's extensive knowledge of humans. Maybe it is too lofty for us, but such a move tells us more about ourselves than the subject matter. We have drawn a line in the sand which respectable researchers are not allowed to cross. Yet it doesn't seem scientific to discount a line of inquiry from the outset. Better perhaps to cast a wider net than simply to utter as a caution, "danger lurks" or "beware the kraken." Who knows, it might eventually earn a place at the scientific and even religious tables.

WAS SCIENCE NATURALLY SELECTED?

According to the Scriptures, God not only knows the feeding requirements of ravens but cares about the lilies of the field (Matthew 6:25–30; Luke 12:22–28). Here is a noticeable difference between natural selection and divine providence, in that the former does not exhibit concern. Natural selection simply waits for the best suited and the most adapted organism to emerge, a type of cream rising to the top, and ensures its success in the form of leaving more offspring than its competitor. It does not lift a finger to assist—and how could it?—for it is passive and not an active agent or force. Those with the most advantageous variations must make themselves felt. Natural selection acts, insofar as we can employ the term, as a sieve, where the successes are separated from the failures. Sieves do not display compassion, but a cold judgment of the in versus the out crowd. The Bible, on the contrary, portrays a divinity who expresses compassion for the world and its creatures and asks us to do the same. Jeremiah 31:20 informs us that God can even be "deeply moved."

But the question must be asked as to the extent of God's care if almost all the species ever having appeared on Earth have met with extinction. Does God's care run out after a time, or does God lose interest in some organisms and turn God's attention to new and improved models? Religion is not generally at odds with science, as we have established, but natural selection does militate against a benevolent deity. Can the two coexist? The answer perhaps may lie in God responding to a cursed creation, that is, attending to a messy world in facilitating the emergence of organisms that can assimilate themselves to it in an ever-increasing way.

God as architect is also artist who paints a greater work of art with every successive brushstroke. But this kind of language is as far as the classical theistic view can take us. From a process perspective, instead, God urges the world on to assume ever greater contours of novelty, which can never be maximized—there is always more room for added measures of it. The process God can utilize natural selection as a tool, ever drawing the world on to creative advancements of the genuinely new. The God of the openness model, regrettably, is still at least partially responsible for the messiness.

Take as a case in point one of my favorite organisms, next to humans, namely dogs. They descended from wolves and exist today in roughly four hundred domesticated breeds. These could be considered separate species, since they cannot all interbreed, if at least for purely mechanical reasons—a Great Dane cannot normally mate unaided with a Chihuahua—but they nevertheless all belong to the species *Canis lupus familiaris*. But natural selection and dog breeding are not the same, for variations which please breeders are qualitatively different from a process that shows no concern whatsoever.

Being best suited to an environment is different from the pleasure dog owners feel for their pets. What pleases dog owners is not necessarily well suited to environments. For instance, Dachshunds are notorious for having back problems, due to their length and short legs. Hence a breed that is less or ill-suited remains, since someone who cares enough about them wants them to be a going concern and their lives, they believe, would be impoverished in their absence. This is not how natural selection works, thus breeders are not selecting naturally but with a view to emotional investment, which natural selection knows nothing about.

Species on the average last for about three to five million years. Certain dog breeds could probably not endure on their own were it not for their adoring masters. However, what then can be said about the attempt at the human experiment on God's part? At the rate we currently assume and the track we find ourselves on, the odds are against us that we will arrive even at the average. Our species, modern *Homo sapiens sapiens*, has been in existence for approximately two hundred thousand years. What are the odds that Las Vegas is offering on our longevity? We might not be around long enough to cash in on our winnings.

Next, a caution against any undue adulation for science, a point to be made by way of an aside. Both my wife and I are sports fans. The more the merrier. We support local teams, but the players do not always appear to support us in return. At times the manner in which they apply their craft seems to transcend the unfortunate into the foolhardy. For them to be unlucky would be to elevate their game. I dare not divulge the identities of these teams, for the mere mention of them makes us grind our teeth and clench our fists and no doubt raises our blood pressure. If we did not know better, we would swear that they intentionally play the way they do just to exasperate us. Our confidence in their abilities has waned and we commit ourselves to put our faith in them no longer. Until next season, of course.

But sport is not the only thing in which faith may be misplaced. Science is to be congratulated for its conquests. It has made a great many strides and we "live long and prosper" more with the advancements it yields. But these do not arise without cost or at least failed attempts, for even science can let us down when seen in the light of examples such as eugenics, asbestos, DDT, thalidomide, and so forth. To declare that science's track record has demonstrated that it can be trusted and for the long term to uncover solutions to current problems is a promissory note and becomes an article of faith—evidently not only the province of religion—and is also ambiguous in its successes. There were and are those who bear the brunt of its fallouts, literally. *Caveat emptor*—let the buyer beware.

Finally, I feel pressed into duty when it comes to the curious decision on the part of some cosmologists to employ evolutionary categories to universes. While true that the universe evolves in the sense that it develops from one overall structure into another and that natural laws can and do sometimes vary slightly from one era to the next, is this saying any more than simply systems undergo change and eventually proceed toward a stable condition as thermodynamics and entropy would have it? And could the description be couched in Darwinian terms? Lee Smolin is one such example who announces that, "The basic hypothesis of cosmological natural selection is that universes reproduce by the creation of new universes inside black holes. Our universe is thus a descendant of another universe, born in one of its black holes, and every black hole in our universe is the seed of a new universe. This is the scenario within which we can apply the principles of natural selection."[4] Really? Aside from the less than conclusive notion that there are universes in addition to our own, one wonders if this is not a stretch, but let's go ahead and make the attempt and see how far we get with the analogy.

We can ask whether universes pass on their advantageous variations due to some competitive and selective pressures imposed upon them by their environment and other entities within it. Precisely which environments would universes have that assess their suitability either to persist or go extinct? And if they were to be more successful than their competitors (are universes really in competition?) by producing more offspring than others, would this pressure cause other universes to meet their demise?. For some universe must then express adaptations so as to outcompete others. Do universes vie for limited resources that would impinge upon them and potentially call for adjustments to be made? Do they struggle for the type of fitness that would help them survive? Are certain ones favored over others for the variations they bear? Do advantageous variations then accumulate so as to produce a new species which could no longer interbreed with its

---

4. Smolin, *Time Reborn*, 124.

parental stock type of universe and produce fertile offspring which thereby would generate a difference in kind?

Does this sound like it could fit within a physics framework? At what point must we confess that the analogy breaks down? And what would prompt an astrophysicist to seek similarity with a biological paradigm? It could be argued that biologists might nurse a physics envy, but this is the first time period we observe in which physicists have bio-envy. Smolin and others like him might maintain that black holes are in fact the sort of resources we speak of and that their number would be the variations needed to distinguish one universe from the next, yet something cannot both be external resources and internal variations at the same time. Instead of being a useful analogy, Smolin's program might reveal a category mistake.

CRANIAL SIGNATURES

Prior to the age of *CSI*, fingerprints were the revolutionary forensic innovation of law enforcement. Uncovering fingerprints or sets of them and checking them against a bank of other sets illumines who might be the culprit who has left them at a crime scene. At least it makes that person a prime suspect. Matters worsen, however, if they do not match any of those contained on file, meaning there may be a newcomer at large who has opted for a career change or has succumbed to an as yet solitary instance in which judgment has failed that person. Worse still if the individual wore gloves and left no prints. Hopefully then some other incriminating evidence was left behind, some calling card on the part of the offender. Later DNA became the evidence of choice.

I have sometimes reflected on what evidence might be produced in the afterlife judgment court to impress upon us the type of lives we have led. The Bible refers to books being opened (Revelation 20:12, 15) containing our list of priors. Yet might there be something else even more indicative of us or undeniable by us? When we undertake what is for us a new decision or behavior and it becomes reinforced and habitual through repetition, our brains take notice, our dendrites in particular. They are parts of our neurons which reach out to other brain cells so as to make synaptic connections with them. Sheldrake would concur and echo that the more we think and act in a certain way, the more these neural pathways will become ingrained or entrenched. This affords the opportunity for chemically electrical signals to travel to the corresponding areas of the body and convey the information they contain, like which neural route to take. Should that route continue to be taken over a previous long-standing one, then a formerly inveterate or deep-rooted neural pathway will give way to another.

This process continues throughout our entire lives. Dendritic pathways incessantly form and reform, and new ones replace old. Our neuroanatomy at any given time is a snapshot of the kind of persons we are, complete with a record of those behaviors we consider a part of ourselves —those acts we are inclined to perform and those choices we tend to make. Given the enormous number of pathways the brain contains, to us one brain looks like many others, but for one who has limitless knowledge of all things past and present, namely the divinity, the feat of reading pathways is achievable. They tell a story and as such are an open book. They can be altered when we adopt a different strategy of reasoning and living, but in the immediacy of the moment, they unmistakably betray what we think and do. They are not indelible, it must be emphasized, and therefore are not a permanent record, unless we remain the same type of persons.

This could be useful to the heavenly court. Brains can provide evidence of "deeds done while in the [old] body" (2 Corinthians 5:10), something we in our new bodies can be shown. Our preferences can be demonstrated undeniably. Now of course there is unlikely to be a heavenly warehouse where brains are kept and retrieved when required in court. Virtual images of the brain, though, could be readily stored and recalled. These images would exhibit what we regard as most important, in essence, our theoretical and practical allegiances. Should we have had a change of heart, the brain would display that as well.

The advantage of the brain's physical structure is that it is unique—no one else has our specific pathways. In that sense they are like fingerprints, yet unlike fingerprints they can be altered through the simple means of adopting a different method of operation. Penmanship in general and signatures in particular can be forged. Speech patterns can be duplicated and the spoken word taped, but may not accurately convey what we truly think, given false testimonies. But brains are a biological window into our actual selves and our hearts and wills are reflected in them. Accordingly, God hopes for changed lives and these changes are recorded in our dendrites. There may very well be a will in addition to a soul/mind, as a mind for some is numerically distinct from a brain, but it leaves its mark in the brain's dendrites. Analogous to the fact that observation of the quantum world interferes with it, since we employ quantum objects like photons to gain information about quantum objects, so the act of investigating our brains alters them, since we use them in the procedure. As fingerprints leave their mark on objects, so do choices/decisions on the brain's structure.

THE PRESENCE OF THE PAST

In an attempt to unveil how Sheldrake can contribute to our investigation, it would be beneficial to clarify how his scheme compares to evolutionary theory. As he outlines, "Morphic fields are shaped by morphic resonance from all similar past systems, and thus contain a cumulative collective memory. Morphic resonance depends on similarity, and is not attenuated by distance in space and time. Morphic fields are local, within and around the systems they organize, but morphic resonance is non-local."[5] More specifically, "Habits alone cannot explain evolution. They are by their very nature conservative. They account for repetition, but not creativity. Evolution must involve a combination of these two processes: through creativity, new patterns of organization arise; those that survive and are repeated become increasingly habitual."[6]

Hence there is an overlap of the two concepts and the language used to describe them. The beginning of morphogenesis is spontaneous; Sheldrake calls it a germ and the corresponding event which natural selection can act upon is known as a mutation. Germs survive through repetition and potentially become habitual; adaptive mutations are preserved though natural selection and potentially occupy a niche long-term. We say potentially because there might be hazards along the way that can derail the process if the new forms or species cannot find a way through. Those germs, having become forms, are inherited by the manner in which morphic resonance molds morphic fields; those genes in the genotype, having become expressed physically in the phenotype, are inherited through reproduction. The force of morphogenetic fields helps to shape and stabilize forms; the passive force of natural selection favors some forms over others and successful ones leave more offspring than their competitors. The two strategies work in tandem.

Sheldrake has been eager to devise experiments that could work toward disclosing the morphogenetic field he has in view. Perhaps he is going about it the wrong way, however. Two of his volumes—*Dogs That Know When Their Owners Are Coming Home* and *The Sense of Being Stared At*—attempt to examine the effect which these fields can have on behavior and

---

5. Sheldrake, *Science Set Free*, 100.
6. Sheldrake, *Science Set Free*, 105.

perception, namely, the ability to detect the intent and activity of another. Unfortunately, his efforts may do little more than create a ripple, if at all, in the scientific community. If his interest is in presenting experiments that, as part of the title of another book of his expresses, "could change the world," then the world at large is paying little attention.

To my thinking, this is because he concentrates on phenomena which themselves have not as yet gained scientific credibility. They focus, rather, on what amounts to ESP—abilities science is unable or unwilling to recognize. This creates a two-pronged difficulty for him, for he endeavors to reinforce the existence of a field which physics has not as yet detected, nor may be capable of detecting, through the activity of human and animal capacities which are more alleged than established, at least in the eyes of those who could render it their official imprimatur. His strategy effectively undercuts the likelihood that he will be heard by those he seeks to reach. Moreover, the scientific response might be that there in fact exists no such field, nor is there any subject matter at all to find per se, much like the proposed "power" involved in telepathy or clairvoyance. It is undetectable because there is nothing to detect, in the expectation of researchers, and those who defend it bear the burden of proof to actually point to something that may not even be out there.

There is another approach, though, that might have a greater impact. Sheldrake, in my view, would be better off tackling an investigation of the kind he has already broached in his writings, having to do with the type of learning that any scientific practitioner would regard as legitimate. He mentioned that if these fields are a reality, then their effect should make it easier for humans, for instance, to learn to operate a vehicle. Over the course of time that automobiles have been in existence, the morphic resonance from those who have successfully learned to drive a car—and judging by the abilities of some drivers in some cities, they are still very much on a learning curve—this should reveal a tendency on the part of new learners to grasp the finer points of operating a vehicle more quickly and effectively.

One drawback to this strategy, already admitted by Sheldrake, is that automobiles might not have been around long enough, now over a century, for a sufficiently noticeable field to have formed. Nor should this shortcoming be castigated, for biologists as well claim that even thousands of years may not be enough time for evolutionary effects to surface. Human lactose tolerance when it comes to the milk from cows appeared about five thousand years ago and took roughly as long in the making after the domestication of cattle in particular and agriculture in general (circa eleven thousand years ago).

Another shortcoming is the unavailability of data from the last one and a quarter century concerning the length of time elapsed for drivers to become proficient at handling their machines. My suggestion suffers from the first but not the second. I propose that a study be undertaken into the alacrity with which newcomers to personal computers are able to master their use, for they can be found globally.

PCs have only been around since the late 1970s, but are they any easier to manage for the neophyte? There are still some for whom frustration levels reach uncomfortable heights and others who perhaps give up too soon. CDs and books on how to work them continue to be produced, so the least we can say is that these efforts are not trouble free. Yet maybe progress can be monitored to determine if the length of time required to reach a comfort level with these electronic devices has diminished. We can gain access to the data and have since observed that the age of their initial usage and mastery has dropped dramatically. If in need, just ask an elementary-schooler.

What we do not have as yet here, though, is the passage of time needed to detect what Sheldrake refers to as a chreode—the contours of the groove that is carved into the background landscape of the morphogenetic field influencing the shape that new forms will take. In this way, there is, in accord with the title of yet another work of his, "the presence of the past" that makes itself felt in the present. Interest in PCs and other gadgets are the germ, but it requires time for them to be reinforced.

This would definitely be a longitudinal study, but as Sheldrake has noticed with his work on rats, those which bear the ability to negotiate a maze pass this trait on to the next generation. What is more, this phenomenon can even affect rats contemporaneously in other parts of the world to pass the same test, meaning distance is not a factor and the "power" (which after all is a descriptor that even Darwin mistakenly rendered to natural selection in his *Origin of Species*) does not fall off either with time or space. Any takers?

This brings us to an extended discussion of both the past and the present.

THE POWER OF THE PAST

There seem to be three main classes of past occurrences that have a bearing on the present. The first is the standard physical constraint that it takes time for an object to yield information about itself. When photons illuminate an object, for example, some of them bounce off the item and hit our retinas, impact our rod and cone cells, create an image in the back of the eye, from which signals are then sent to the brain. This allows us to perceive the object and form an impression of it.

The time interval between photons bouncing off an object and hitting our retinas means that the information captured is in the past once we perceive it. No longer is it the same object, for it might have changed in the interim. This is most acutely the case with objects at astronomical distances, for there is no telling if faraway stars, for instance, from which light may require on the order of billions of years to traverse the space in between, are still as they appear or if they have undergone a significant alteration or succumbed to another fate, such as having gone nova.

The second way the past can impact the present is through our recollection of it. Our memory of an event in this case is a present interpretation of a past occurrence. This means not only that accuracy of the event can become modified over time, but that our recollection is in the present, despite its having a past referent. Perhaps, then, it would be more precise to intimate that we are being struck by a present impression of a previous situation. As such, it is not strictly the past which is having an effect.

In the first case, the source of information is necessarily restricted to the past, assuming the limitation of the speed of light. In the second, however, our mental faculties recall a previously current but now past event and we relate to it. If the recollection is that which is being responded to, then it seems as though both are occurring in the present, though the memory of the past can lack precision and become distorted with time. In a similar way to how dendritic connections are plastic and can change connections throughout our lifetimes, so there is a plasticity of memories, in that they too are filtered by the passage of time and personal predilection. We can relive a moment which packs a punch, but the moment is past and the punch is present. It packs a punch now because it may have already packed a punch then and we may react to it differently the next time around.

This brings us to the third class. In addition to the present and the immediate past, we might also stand in relation to the more distant past. We all come forth from our mothers and this event is in the past for all of us—some more so than others. The vast majority of persons fail to have a recollection of their entrance into the outside world, but we all stand in relation to it. Whatever and how far back our earliest memories, they usually do not extend to our first post-partum events. Our inability to recall our birth occurrence does not affect our status as offspring, nor does it drop off with time. We continue to be sons and daughters regardless of age and memory.

Yet here is a situation in which the past truly impinges upon the present in a way that the passage of time cannot alter. Birth sets us on a trajectory that has but one initial reference point (setting aside for the moment the possibility of past lives). No amount of looking to the immediate past will help us completely explain our present status as offspring. The more remote past is required to assist us in this respect, a past we do not normally reach, for memory does not take us there. We have a conceptual grasp of it—we must have had a birth—and might even point to a visual record of it in the way of photos and documents. But it resides in the imagination in ways that go beyond this kind of evidence. Recall it or not, we are impacted by it. The past is unavoidable.

Now which of the above three types can properly be referred to as the causality of the past? In the first case, the past as such is not causal, but only the present, owing to the duration of time required for information transfer. The past has gone. What, though, of the circumstance for close proximity of subject and object? We mentioned one extreme having to do with distant celestial bodies and events. What about the other extreme? Say we bump into a chair in the dark and stub our toe. According to neurophysiology, the peripheral nervous system (PNS) will initially register the pain as a sharp one and only slightly later will the central nervous system (CNS) process it and perceive it as a dull and protracted one. Time is required for both to take effect, even though the toe will be affected immediately, even prior to our ability to register it. A part of our body is traumatized and then we relate to it. Response times are delayed, yet the event has occurred and not without anatomical and physiological effect.

In the second case, memories also exert efficient causation, though the memory itself is present cognitively and perhaps physiologically, depending upon how we relate to it, albeit the event from which it derives is in the past. When I remember something, it is my present self doing so, shaping the original occurrence as I go. I may react differently to it than I did initially.

The passage of time together with our reflections on the event can modify the way we think and feel about it, to the point where the memory itself is altered. We may remember it in ways that do not accurately reflect the original event. Our imaginations might be more active than our powers of recollection. Memories can be chiseled for contemporary needs. Issues of trustworthiness of witnesses plague the courts, even if without conscious malicious intent on their part. Our capacity to recall accurately can be open to suggestion. This is both the boon and the bane of brains (pardon the alliteration). If brains can be plastic, so too can versions of the past.

In the third instance, who we are is a direct result of our past. Even if we were to change our names and undergo gender reconstructive surgery, we remain offspring of our biological parents. No amount of renovation in the intervening time can alter this status during our earthly existence. At least here we can actually point to an event as being efficiently causal in a non-trivial way. And this is independent of our immediate past in addition to our memories. Law enforcement agencies might label unidentified bodies as John or Jane Doe, but even these persons are effects of their parents, whoever they may be.

Given the above cogitations, is it essential to consider the past as imparting efficient causality? Whiteheadians insist on it, for their metaphysical system requires it. Present experiencing subjects for themselves become past felt objects for others. For Whitehead's followers, this constitutes the extent of data to be obtained for any and all subsequent rounds of becoming. Is this overstating the case? Does Whitehead, in an effort to avoid the fallacy of misplaced concreteness, commit the fallacy of hypostatization and make a concept, in this case the past, something it is not? Is the idea that abstract subjective processes of concretion leading to concrete past objects also misplaced? How objective are the objects in the past? For Whitehead they are a given (though Immanuel Kant, for whom nothing is given, save for this statement, would not concur). They impact us viscerally before being processed cognitively. Are Whitehead's efforts successful in translating the past from the level of inaccessible subjective concept to accessible realm of objects? Or does the past remain a concept, bereft of anything concrete?

Or, more to the point, is the past a reality, and if so, where is it? Are we referring to something which bears no ontological existence? And can something which can be described in this way even exert efficient causality? We are stuck with our own biographies, but does this objectify the past? The past cannot be changed, but our attitudes toward it can. Does this constitute an effect that the past can boast? We are counselled to live and stay in the present, but where does the present go once it slips into the past? Relativistically, all objects are in the past because it takes time for information about them to reach us. Hence the past impinges on us, but does this make it real and something more than simply the present when it actually becomes felt? Does it imply that the past and present coexist? It can affect us, but not we it, except for altering it to suit our more recent purposes. Since there is an arrow of time and if there is efficient causality, then the past can be the only vector of it, the actual effect appears to be unidirectional.

In addition to memory distortion, archivists are aware that documents and photographs deteriorate with time; and even though they come from the past, they no longer look like they did when first produced. Items from the past are not past items; they are in the present; they carry and convey information from the past, but they need to be in the present in order to do so. Thus pasts differ in relation to us, as relativity informs us. But the present is all we have to work with; with it we can alter the way we view the past, but this is not changing the past. The past is permanent and settled, but not when something from it enters the present. At that point, it must succumb to the distorting effects of the present.

Efficient causality cannot arise from that which does not exist. A push can originate from an event that is past in relation to the recipient, like a memory, but the push is present and the past is no more. One cannot point to either past or future. So what are we left with? The past as something not having reality shapes the present, but the present as something having reality contains that which was from a past we cannot reach; thanks to relativity, it reaches us. Yet the past that reaches us is intangible; it is information only. Once in the present the impact is felt. The past which can be malleable is not the actual past, but the past is not actual anyway. For the tactile fans among us, this is not completely satisfying. But that's the past and present for you: not always delivering on their anticipated promises. I imagine we must lower our expectations.

Paradoxically, the past is forever beyond our reach, but it makes itself felt, as both Whitehead and Sheldrake would urge. In order to have an impact, it must interface with the present, even if it never gets beyond the conceptual. Memory and imagination are insufficient to ontologize (actualize or make real, something having being) or objectify the past, but that by itself does not prevent it from packing a punch. There cannot be objects as such in that which is abstract, but only objects impel the present. The past must be real and unreal at the same time. Whiteheadians would come to the rescue and announce that internal relatedness is the solution. We are impacted by our felt import of our past for us. We viscerally receive from it and respond presently to it. In this way, subjects like us in the present become objects for both ourselves and others in subsequent rounds of becoming.

If becoming is pre-eminent over being, then experience as a metaphysical category solves the conundrum.

We think too materially about objects, they would declare, for theirs is not a substance view of reality. Why, then, do they refer to the contents of the past as being objects? If their intent is not to confuse the material with the immaterial, then they will need to search for another term, for there cannot be objects where there is no substance. In that case, their whereabouts would not be in the past, for they are ever present for instantiation into our worldly realm. This is not what Whiteheadians intend. What Platonists reject, however, is flux, since it would be unbecoming (no pun intended) for the world, because for them change implies corruptibility. Yet flux is precisely that which Whiteheadians embrace; in fact, it is an essential ingredient in their system, much to the chagrin of Platonists. Hence Whitehead is only peripherally Platonic, for there is a lot of flux around! The abstract becomes the concrete, all the while denying that there is anything substantial that we can point to as concrete. Whiteheadians will need to invent a term other than objects as the constituents of the past. When asked where these objects reside, a process thinker at a conference I attended confidently responded with "in the past where they belong." This is an unsatisfactory answer. It is elusive and evasive. Please try again.

Next I wish to focus our attention on three themes: free will, feelings, and emotions, respectively, all topics important for process thought.

LET MY PEOPLE CHOOSE

Just a short statement on this subject. Charles Hartshorne (1897–2000), another process thinker and teaching assistant for Whitehead while at Harvard, understands the freedom of entities to be real but not radical. We possess greater self-determining power than any other creature, save for angelic beings, at least to the extent that process thinkers hold to their existence (God is not a creature as we saw in part 1). Despite this considerable amount of free will, we are also ineluctably determined. We are bound by our past and physicality, meaning our freedom is actual but not extensive. We have the capacity to reject God's initial aims for us, and to say no to the divine prompting is a significant amount of freedom. We can trump God's will. Once again, this is real but not radical freedom. We cannot be other than human and mortal in this life, nor can we lay claim to eternity as God can. Nevertheless, we will inevitably insist on our own will at certain points, at which time some measure of disorder enters the world, which some call evil. Freedom can engender bondage to destructive elements, which no longer looks like freedom.

While authentic freedom may be found in adopting God's purposes, real free will is also exercised, given the human condition, in ways that militate against God's will. We are constrained by humanness and so cannot hope to approach the type of radical freedom which God enjoys, yet even the real freedom of the Christlike kind is also unattainable for Earth-dwellers in our current state of alienation from God. But with the empowerment of God, we can pursue it. Real freedom is that to which humans can aspire and is that which also occurs given our feeble attempts at being Christlike. Entities of all types employ some measure of self-determining power, enabling them to synthesize their data in spontaneous ways, thereby genuinely transcending their own inherited past and contributing to the creative advance toward novelty. This is real but not radical freedom; the latter is forever beyond our reach. That should not trouble us, for we are not built for the radical kind.

IT PAINS ME TO SAY THIS, BUT I FEEL I MUST

Pain, we are told, is an indicator of a condition that requires attention—a biological warning sign that an organism's well-being might be on the brink of a threatening situation, and an adaptive sign at that, selected for as it is by natural selection. We would like to avoid it, yet it works for us, for without it we would have fewer signs to heed so as to recognize when we are in a vulnerable position and our system is under attack. There are, of course, other signs such as illness, which communicate to us the need to address a circumstance that has made us unwell.

Feelings, an important category for process thinkers, are also indicators of internal conditions. Certain feelings betray specific subjective states. This subjectivity makes feeling a less than measurable category, especially for students of positivistic science. The two of them, pain and feelings, are indicators of organismic health or harm, but only the former is deemed a legitimate health issue, even though it can only be subjectively and not objectively quantified. Health practitioners might ask us to declare on a scale of one to ten, ten being the highest, how painful a certain sensation feels, but this hardly amounts to a controlled scientific laboratory experiment. While a branch of medicine is devoted to emotions, namely psychiatry, even there a biochemical basis for such feelings is sought, almost like a knee-jerk reaction. This is regrettable, for the scientific community in general and the medical in particular, in addition to other interested parties, could benefit from, say, cataloguing feelings and the concerns to which they point. Perhaps some preparatory work can be undertaken for just such an endeavor.

At the very least, feelings are noticeable in us as well as in others. We recognize them when we have them. They do not surface in discrete packets, nor can units be applied to them, yet their severity can be assessed. There are times when more or less of a feeling may be present, and despite their lack of quantifiability, multiple feelings may arise concurrently which are distinguishable and possibly separable—they may or may not overlap. Anticipation, for example, might be combined with fear of the unknown. Feelings are unmistakable in that no amount of prompting us otherwise can convince us that, for example, the joy we are experiencing is actually shame in disguise.

So far there is little controversial in this analysis. Even positivistic scientists (not all scientists find themselves in this fold) could nod in agreement. But the upshot is that pain continues to be an accepted term in medical circles even though it also lacks the discreteness and unit assessment that feelings also do. Jeremy Bentham was unsuccessful in assigning units to his pleasure and pain calculus, but this does not deter the medical community from taking pain seriously. "Pain is an adaptive Darwinian trait for organisms," they might be heard to chime, yet feelings may yield a similar adaptive strategy.

But we are getting ahead of ourselves. Back for the time being to issues surrounding the characteristics of feelings. Few deny the presence of feelings in their lives, and in this sense subjectivity is pervasive throughout the human world—objective in so far as we can all lay claim to it, save for the comatose. We all possess them in spite of some persons being described as unfeeling. Feelings appear to operate viscerally and their effects may surface in different parts of the body, like our hair may stand on end, we might break out into a sweat (cold or otherwise), and various areas may tingle. But when we claim to have feelings, they seem to be contained in the internal organs, somewhere between the head, which can throb and ache; to the throat, which might have a lump in it; to the lungs, which can be regarded as tightened; to the stomach, which can churn in anxiety; to the heart, which can grow fond or be broken; to the intestines, which can have hunches and thus gut feelings; to the bladder and colon, which can empty themselves when faced with what is perceived as life-threatening situations. Hindus also refer to the chakra—a point midway between the anus and the genitals which is sensitized in at least heightened if not also life-or-death circumstances. Hence subjective states do have their effects at the organ level.

We even attempt to externalize certain feelings in artistic expressions. The images that hate, anger, and rage conjure up include redness, perhaps because our faces can become flushed when we experience them. But the feeling of embarrassment can produce similar colorization while the facial expression alters. And seemingly the feelings we have can further be manufactured by us—we can decide to be happy or sad. There need not be a one-to-one correspondence between events encountered and response emotions. The same event can elicit various reactions in different persons and at different times.

Even if the same emotion were to be felt, the intensity might vary with each experiencer. If beauty really is in the eye of the beholder, then the feeling of awe when encountering this beauty will depend on the extent to which the beholder takes the object of his or her awe as beautiful. We cannot be certain that, say, what one individual feels as guilt has the same characteristics as what someone else feels. Each may also bear a roughly equivalent amount of guilt, but the effect of it or the impact it has may differ from one person to the next. The same with pleasure and pain, which has ramifications for the penal system. One can withstand more of the latter than another, so should more of it be applied to him or her? As Morris Berman admits, "one reason that I am groping in the dark is that we do not *have* methodologies of feeling, only ones of analyzing."[7]

Perhaps assembling an inventory of feelings and cataloguing them might prove to be helpful personally and medically, despite their being unquantifiable. In line with Whitehead's insight, much of our lives hinge upon them and we sometimes also permit them to be the ultimate factor in our decision-making processes. Even the otherwise rational among us can be caught doing so.

---

7. Berman, *Coming to Our Senses*, 131. Italics original.

CHANGES OF MIND

There was a time when I was well-nigh a card-carrying member of the process theological fraternity. Later I became a friendly critic. Now I am just a critic, though an appreciative one. Whitehead, I contend, has made a significant contribution to philosophical and theological thought, one that ought not to be skirted around but should be met with head on. In my view, if advancements are to be made in these fields, it will be by going through Whitehead, not in bypassing or sidestepping him. Nevertheless, there are difficulties with his strategy. I will list some advantages and disadvantages of his scheme, with the latter, as I see it, outnumbering the former.

First, on the plus side, Whitehead builds a bridge between science and religion and his view holds much promise in the way of explanatory power with science, in that his approach weds physics and metaphysics. Further, philosophical problems with the divine metaphysical attributes, such as omniscience and omnipotence, are overcome, for God neither knows the future with certainty nor possesses the power to enact it. And the objection that without power in the usual understanding God cannot be an agent can be countered with the appeal that one should not underestimate the efficacy of suggestion.

Next, the process God is absolved from the charge of arbitrariness as it can be leveled in more interventionist proposals, since Whitehead's God operates only through persuasion. Moreover, in terms of the impetus which ignited Whitehead's attempt in the first place, it provides an answer to the problem of evil, known as the theodicy problem, since God is divested of the responsibility for it. Instead, it rests entirely on us as those who can either accept or reject God's ideal for us that God hopes we will grasp and embrace. These advantages are few in number yet not in import, for they by themselves would constitute a major boon to what persist as perennial concerns.

But this is not all. Initially, on the minus side, it strains credulity to imagine that subatomic particles like neutrinos enjoy interiority and exercise at least an iota of self-determining power, which they must since they and all entities find themselves on the process hierarchical scale. All occasions of experience reside somewhere on this scale and thus cannot be without the capacity to both feel and respond. Theory here, however, might not correspond to reality. Second, contra quantum theory, there is more indeterminacy as one goes up this hierarchical scale. In quantum mechanics, elementary particles are those about which the least can be known with certainty in terms of specific characteristics over against others. They are indirectly proportional in the sense that as more of one property is known, such as position or momentum, the less of another can be unlocked simultaneously. For process, though, humans are those creatures about whom God can have the least amount of certainty.

Third, God cannot inherit evil, since evil is of no value and God can only inherit that which contains value. Yet evil is in the world, so how can God avoid it, since what we do with God's purposes together with our free will is the world which God inherits? Where then does evil go? Nor is there any assurance that evil will ever be eradicated. Fourth, in line with the previous point, there is no final court of appeal on offer to redress grievances and correct injustices outside of humans, and we are not always good at it. Plus, why do we insist on God being persuasive but law enforcement coercive? This seems to imply that there is a place for coercion after all. Apparently it is acceptable though not ideal for humans, but God is above that sort of thing (at least in terms of ability if not also volition).

Fifth, since God is the total reality and therefore has no external environment, God cannot be in heaven or any other place for that matter, as there is no place outside of God, meaning heaven (and hell?) must be internal to God. Sixth, even the process God is Greek when it comes to the static, unchanging nature of God's primordial aspect, the inexhaustible reservoir of divine ideals and potentialities. These do not undergo any alteration and as such the greatest proportion of God by far is static. Seventh, there are two afterlife possibilities and neither option seems appealing. Objective immortality, on the one hand, means that our lives are simply monuments in the mind of God, while subjective immortality, on the other, will not work since something like consciousness must survive so as to exhibit subjectivity on our part. And if so, there must in fact exist something other than actual entities, which is contrary to Whitehead's ontological principle that the world is exclusively and entirely composed of momentary occasions of experience. But immortality is not momentary. The eighth follows upon the seventh, in that if there is indeed a continuation of awareness, then this sounds conspicuously like a soul to me, entailing a dualistic approach after all, despite the effort to supersede it.

By my reckoning, there is more against Whitehead's treatment than for it. As I reflect on it, and reintroducing a previous topic, God cannot be coercive because all action is from the past, yet only the past is that which can actually yield efficient causation in the process system. To the extent that the past wields or packs a punch, it bears an impact which we cannot ignore and imposes upon us some type of response. As a result, then, the Whiteheadian strategy is not very strategic if it downplays coercion but is built on it. If God uses the past to make an impression on us, as any past event might (some more so than others), then coercion becomes a necessary feature in the scheme, for respond to it somehow we must. One cannot have it both ways without a theoretical overhaul. Still, having said all this, and to reiterate, it remains that one should not underestimate the power of suggestion, for impressions can be impactful.

But his approach was not the only one we examined. Recall that Sheldrake's view is that "[e]ach species has a kind of collective memory. Each individual both taps into this collective memory and contributes to it."[8] This sounds strikingly similar to the Whiteheadian concept and language of "the many become one and are increased by one,"[9] as well as "What had been a subject for itself is now an object for others."[10] Sheldrake further compares brains, which he claims do not necessarily store memories, to TV receivers, which "tune into invisible resonances . . . transmitted through invisible radio waves."[11] If invisibility were the only concern here, then the difficulty I find exposed could theoretically be overcome, yet the difference with TVs—and this is why I have not fully signed on to Sheldrake's potentially revolutionary work—is that their waves and resonances can be detected by electronic equipment, whereas his cannot. Being undetectable leaves them in the ranks of the speculative, and I am not entirely certain that this situation can be remedied. Reality is not compelled to yield to our preferences.

This is not all. Both Whitehead and Sheldrake would concur that memory is one instance in which something can be efficiently causal without being material, since it often demands a response despite its being undetectable. We might know at which point a person experiences a memory when a certain cerebral component lights up in electroencephalograms (EEGs). What we do not have is access to the actual memory, since its contents are undetectable by anyone other than the experiencer. More accurately, what our memories mean to us is a private matter, ours and ours alone. Nor would we at the same time conclude that there was no memory recalled given our inability to decipher one. Brain centers can signal when one occurs, just like a red light signals when radio announcers are on the air,

---

8. Sheldrake and Shermer, *Arguing Science*, 76.
9. Whitehead, *Process and Reality*, 21.
10. Griffin, *Reenchantment of Science*, 154.
11. Sheldrake and Shermer, *Arguing Science*, 77–78.

but the message escapes us unless we tune in to the station. Consequently, it is not memories as such that are detected but their effect on the brain, and their nature is kept from us unless the experiencer divulges them to us. At that point, though, issues of reliability arise. Hence we can tell when but not what, the occurrence but not the significance or subjective import.

Sheldrake's hypothesis appears to be in the same boat. We have access to an alleged effect, namely form or behavior, but not its cause. By this reasoning, Sheldrake's fields are not undermined for being undetectable. They and memories can still occur. Morphic fields explain what genes cannot, or more accurately, morphic fields take over where genes end. Specifically, our body's cells and molecules are replaced continuously and according to Sheldrake genes are ill-equipped to navigate where the new ones should go, a process known as regulation, because they mostly code for amino acids, which combine into polypeptide chains named proteins, and little more.

As for the capacity of genes to direct this traffic—in Sheldrake's view not appreciably—geneticists are uncovering regulator genes that can do the job, switching genes on and off where warranted. Sheldrake proclaims that genes are incapable of shuttling cells to their destination as, say, arm or leg cells; geneticists claim they are beginning to understand how genes in fact could accomplish the feat. This places Sheldrake's notion in jeopardy and becomes another reason why the biological community views it in negative terms. All I can counsel for interested parties is to stay tuned for any forthcoming developments.

While true that Sheldrake's still useful offering, like Whitehead's, boasts explanatory power, let's tally up the most important items. On the plus side, as mentioned, whereas most of the molecules in our body, DNA excepted, experience a turnover within a range of days to months, self-resonance explains how the form of an organism is sustained despite this incessant molecular replacement. Moreover, at the point of death, the conscious self together with its memories could remain intact through being intimately connected to (even absorbed by?) the morphic fields which helped to give rise to them. On the minus side, these could just as readily dissolve into the field and be lost forever. There is no guarantee that the conscious self and its memories would remain intact, for fields by their very nature are not discrete. But the most telling shortcoming of Sheldrake's strategy, as intimated above, is the lack of detectability of his fields. To my thinking, the impartation of information requires some energy, despite Sheldrake's insistence that his program is non-energetic. Resonance is energetic, for it transfers energy through the propagation of waves, like a tuning fork can shatter a glass. So his fields ought to be detectable.

For both Whitehead and Sheldrake, then, I imagine I should self-declare as an arm's-length suspicious appreciator. I have undergone several modifications theoretically and practically throughout the course of my (gratefully) unfinished history. So in what sense am I the same or a different person? Part 3 will attempt to address this and other concerns. As a final instalment before we embark on that journey, I wanted to point out a misunderstanding. It should be noted that chance and time are often referred to in evolutionary theory as factors in the process, but properly speaking neither of them is or can be a cause, as if to say that given random activity and sufficient time these ingredients will produce something when fully baked. But they have no power to create, nor do they hold agency status. Instead, they are the arena for opportunities in that they allow for the arising of certain possibilities in natural selection and no more.

PART 3

Charting Our Course

WHO IS GOD?

Continuing the methodology from parts 1 and 2 of progressing from biblical to philosophical-theological topics and then scientific, I offer the following as an initial biblical theme. As alluded to above, there are classically understood to be approximately nine metaphysical attributes for the Judeo-Christian deity; there are also roughly eleven moral properties, we being able to participate only in the latter. This, of course, does not exhaust the descriptive possibilities or present a complete portrayal of the contours of divinity, and nothing ever could.

Nevertheless, we make the attempt because some humans throughout history have been subjected to direct religious experiences, which might disclose what God is like. These are primary experiences which become secondary once committed to writing, given the limitations of both language and reporting. Others who come later then interpret and draw implications from these documents in a tertiary sense. And so it continues at additional removes for each step. This should not deter us from making a stab at the project with the unique perspectives that we bring to it, for no one else entirely enjoys our particular vantage point, and perhaps, with God's illumination, this can afford fresh insights. Hopefully what lies below can add to the discussion.

If one were to ask what category or genre of literature the Bible falls into, one could respond thusly: since God is both revealer of God's nature and purposes together with their concealer, for we will not know fully until the end (1 Corinthians 13:12), this describes a mystery. My wife enjoys a good mystery novel and reads as many as she can get her hands on, and we both view an assortment of mystery series and films. Mystery differs from suspense in that for the former whodunit is kept from the audience and is revealed only near the end, and if successfully done will engender a range of responses from "I figured it was him/her/them" to "I did not see that coming." As for the latter, the assailant may already be known to the audience, so the suspense comes in, say, whether the antagonist will intercept the protagonist before s/he can reach the authorities with information about the offender. The viewer is in suspense to discover whether s/he will arrive in time against all odds. Since most of these art forms have a resolution that is "happy," the protagonist gets rescued and the antagonist gets caught.

If we have consulted the Bible from cover to cover, we already know the ending, for God wins the battle against God's enemies and makes a new home for God's allies. So the issue is not whether we are in suspense about God triumphing in the end, for we are already informed and have viewed it textually along the way. What remains is the undisclosed identities of figures such as the Son of Man (is he the very same as the Messiah or a different person?), the Accuser (does he really have horns?) and angels (do they really have wings?). Plus, what are of inestimable concern: How young or old will we be in the afterlife? Will those who died in infancy be there? Will there be representatives of other human species such as *Homo erectus* or Neanderthals? And will not only loved ones but family pets be there as well? Not to mention what God really looks like. This information is kept from us and will be disclosed at the end, and that makes the Bible an unsolved mystery.

Three terms are employed to describe this state of affairs. The first, understandably, is *mystery*. Paul mentions that he is about to tell us a mystery (1 Corinthians 15:51) and offers a glimpse into how the resurrection body differs from our current one, and he can only do so with the limitations of language and must resort to analogy. As there is both continuity and discontinuity between a seed and the plant which the seed will germinate and grow into, so will it be with the bodies in the life to come (verses 35–38). Likewise, there are mysteries regarding the will or plans of divinity (Job 11:7; Ephesians 1:9; Revelation 10:7) as well as of Christ and the gospel (Ephesians 6:19; Colossians 2:2; 4:3).

The second term is *hidden* and is sometimes incorporated into the first, for it is claimed that the mystery of the gospel—that God has opened a way for the Gentiles to be accorded the same privilege of kingdom citizenship as the children of Israel—was kept hidden for long ages until the appearance of the Messiah. It had been made known in the Babylonian captivity that the God of Israel is a universal and not just a tribal deity—a God for "all nations" (Ephesians 3:3–9; Colossians 1:26; Romans 11:25; 16:25–26), and it is Christ "in whom are hidden all the treasures of wisdom and knowledge" (Colossians 2:2). God also reveals mysteries and hidden things, thereby enabling, for instance, the prophet Daniel to interpret the meaning of the dreams of multiple rulers in Babylon (Daniel 2:18–30, 47).

The third term is *secrets*, say, about God's wisdom, which has been entrusted to God's apostles and to be revealed in due course (1 Corinthians 2:7; 4:1). The disciples of Jesus were also "given to know the secrets of the kingdom of God/heaven" (Matthew 13:11; Mark 4:11; Luke 8:10). Finally, with respect to the nature of God can be added a fourth term. In accordance with the term hidden, God is described as one who not only "hides himself" (Isaiah 45:15) but also dwells in sometimes "thick darkness" (Exodus 20:21) or a "dark cloud" (1 Kings 8:12). In addition to living in "unapproachable light" (1 Timothy 6:16), God is secretive, hidden, and paradoxically surrounded by darkness. In fact, paradoxical might be the best and safest way to describe God, since God confounds our puny rationality. And this may be the point—we cannot bend our heads around God because God is greater than the extent to which our thinking, language, and imagination can take us.

Anselm, an eleventh-century bishop of Canterbury, was on to something, but only in part, when in his ontological proof for the existence of God he announced that God is "a being than which nothing greater can be conceived." More specifically, even our conceptions of God pale before whatever God's true nature is like, and all we can hope to achieve is to piece together a few clues about what God has revealed to us or left for us to uncover. Anything beyond this is unclear, given God's status as having "gone dark"—still active, though in the shadows. Best, perhaps, to portray God as a secret agent whose particulars are classified, disclosed only on a need-to-know basis.

As for my recommendation and as far as mysteries go, the Bible is a pretty good read.

"[S]O THAT GOD MAY BE ALL IN ALL"
(1 CORINTHIANS 15:28)

Regrettably, license has been taken with this passage to declare what it was not intended to convey. There are those who suppose that what is meant by these words—and here we reintroduce process themes—is that at the end of history all of creation will be absorbed into the Godhead. This will result in God essentially switching gears from either a classical or openness posture, where God is more transcendent than immanent, to a panentheistic format (a process term meaning everything is in God), in which both God is in the world and the world is in God, as Whitehead claimed.

This position, sadly, has not been worked out theologically or philosophically. God does not take on the mantle of one type of deity only to shed it at a later time, nor does what is inappropriate for God in the present somehow receive approbation at the end. In bald terms, this strategy runs counter to the divine metaphysical attribute of immutability, which the classical camp upholds, in which God undergoes no change. I am at a loss to envision how this would not constitute a change, and a major one at that. The very idea that there is currently a subject-object distinction between us and others, including God, in the classical and openness senses, and that subsequently this will give way to a diminution of such distinction, is a head-scratcher.

There definitely are points of contact between the classical God and the world this deity created, such as the sacraments and the incarnation, yet to suggest that this gives way to a type of cosmic eschatological equivalent to a microbial phagocytosis, in which an amoeba, for example, ingests a food source by incorporating it through enveloping it, is not the intention of the biblical passage. God does not become one kind of divinity and then metaphysically another when it suits God. God is either transcendent with limited immanence or fully immanent in God's concrete actuality while at the same time transcendent in God's eternal essence. To announce that God is the former while there is a history to speak of and next the latter after this same history comes to a close is to play fast and loose theoretically and to confuse categories.

The panentheistic divinity does not enjoy the same predicates as does the classical deity; in fact, the process God has none of these attributes as classically understood. For God to bear these properties initially and then to forfeit or reconfigure them eschatologically is to promote a smorgasbord of attributes, some more or less palatable in a given era or epoch. The classical divinity does not become panentheistic "when the rent comes due," for metaphysical attributes are not so easily divested. The notion that God, say, has all the power at one time but lacks it at another is irrational. Rather, the passage intends that in the end God will be everything to everyone, where God's unrivaled rule will become fully manifest, God's kingdom will be universal, a new order will be established, and every need fulfilled. Those are the changes we can expect.

IT'S ONLY A MATTER OF TIME

Unsolved problems persist in human thinking, whether viewed from the scientific or philosophical perspectives. They are referred to as the big questions by science and limit questions by some philosophers of science. These issues are beyond science's competency in particular and human understanding in general to unravel. The conundrum is whether these concerns are unanswerable in practice or in principle; that is, will they ever be resolved or are they forever beyond our reach? Those with faith in technology estimate that it is only a matter of time before we will conquer the previously thought unconquerable. I am not so confident. Whitehead speaks of the fallacy of misplaced concreteness; there may also be a fallacy of misplaced confidence. These issues involve, inter alia (among other things), the beginnings of the universe, life, and consciousness. The following timeline will situate the difficulties.

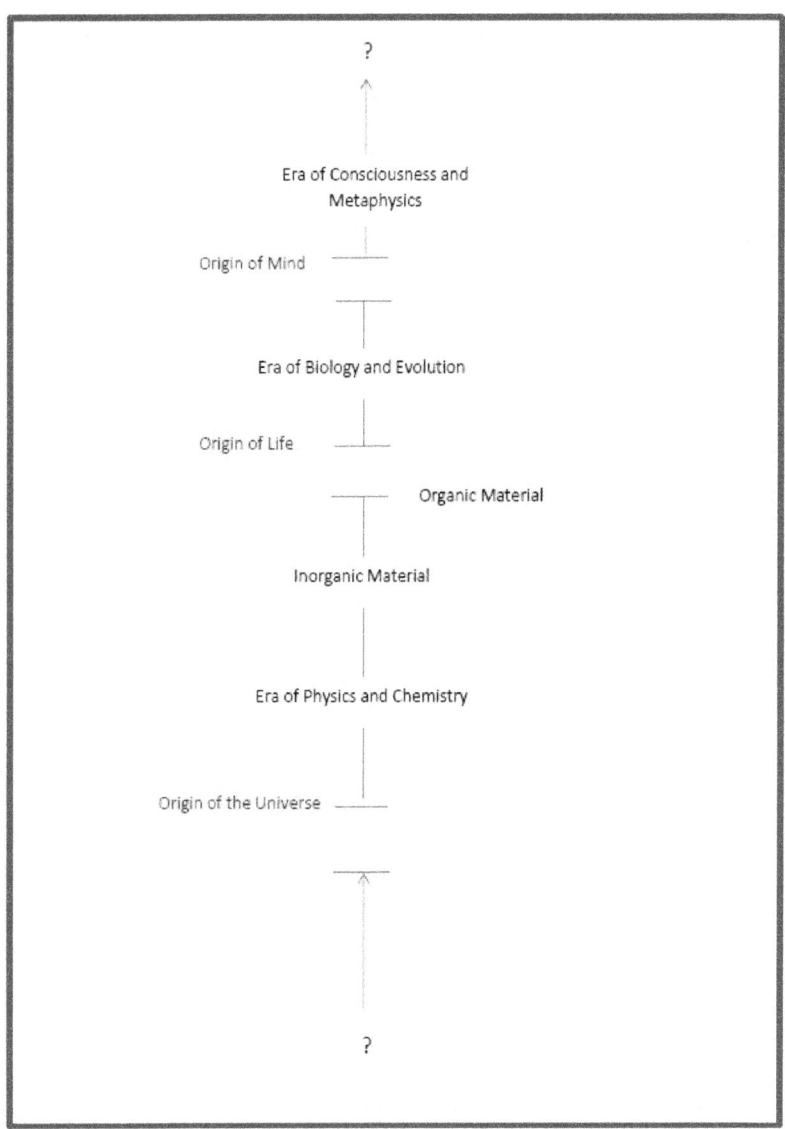

Illustration 1: Timeline of Universal History Leading to Us

At one point there was not a universe and at another there was, or as some physicists will tell us, this is to confuse matters, since time itself is a product of the beginning of the universe. To have space is to have time, and to have spatial geometry is to have space-time curvature and thus gravity.

No universe can be without it. Yet the question remains as to whether there is another dimension or another universe that gave rise to our four-dimensional one. And if so, then how? In any case, the beginning of the universe yielded the elementary particles, some of which eventually cooled and coalesced into atoms. As matter coalesced into galaxies, after about one billion years, additional elements were cooked in the centers of large stars. Then the remaining naturally occurring ones appeared after these large stars went nova or supernova. (These comprise the elements in the periodic table up to and including uranium; those beyond it are synthetcally made in the laboratory and are radioactive.) This was the era of physics, when all the elements came on stage.

Later, when planets, asteroids, comets, and meteors, and hence solar systems formed, atoms combined to produce molecules and thrust us into the era of chemistry, physics having had a sizeable head start (perhaps why physics sometimes displays a superiority complex, for it was first on the scene, and why other disciplines may exhibit physics envy). The universe has since cooled sufficiently so as to be bathed not in the visible light portion of the electromagnetic spectrum of radiation but in the cooler microwave region, giving us the microwave background radiation or MBR. Eventually inorganic material gave rise to organic, paving the way for life to appear, though it is important to recognize that no amount of inorganic material will give you organic, and there is controversy over whether a critical mass of brain matter will get you mind.

But not so fast, for there is another enigma as to how life surfaced where there was none before. No one, outside of the fictional Dr. Frankenstein, has been able to fashion the living from the non-living. Nevertheless, once life arose it seemed to thrive against all odds and found additional ways and means to flourish. We then entered the era of biology and the evolution that carried it along to greater forms and numbers. Most of these perished long ago. Unicellular organisms became multicellular, though most of the biomass of the biosphere is still unicellular, and vertebrates emerged along with their central nervous systems (intending no disrespect to those invertebrates, like the octopus and squid, that would object to this vertebratist superiority and exclusivity).

Then another event occurred: some of these forms expressed the faculty of and ushered us into the era of—call it what you like—awareness, consciousness, or mind. These also arose where there was none before, and none of our concoctions can duplicate it as yet (although some artificial intelligence [A.I.] designers might be hopeful). Hence it can be coined the era of metaphysics. It is difficult to decide how far down from humans on the hierarchical scale mentality resides; perhaps this is more the jurisdiction of speculative philosophy. But there they are—three enigmas which confound our best efforts at closing or disentangling. How did we pass from one to the next? Not our universe to our universe? Not living to living? Not thinking to thinking? Not reflecting to reflecting? Not philosophizing to philosophizing? Do we expect that technology will eventually bring the unknown to light or solve these mysteries? Or is science ill-equipped to perform the feat? Plus, is ours the final stage of eras or will the present one yield another, even greater one?

It might be presumptuous for us to say that we are the culmination of evolutionary history, especially given our penchant for being deleterious to what already exists. The efforts we extend, if they are injurious to the natural order, will eventually recoil on us. If so, then we may actually be the culmination by default if we precipitate the demise of life forms. At that point, Earth will not require an extraterrestrial hit so as to take the evolutionary process along a different course. We could manage this all on our own by mismanaging the Earth. Alternatively, that could allow a different set of organisms a turn at being sapiential and perhaps do a better job than we did.

I wonder if they would then have the same preoccupation with these limit questions as we do, or if their inquiring minds will take another path. I hope they will not make the same mistakes. Maybe they can decipher our fossilized languages and learn from our own history as to what to avoid. We had better leave behind a time capsule for them to find, and should the surviving form be A.I., then we best leave them a copy of "What Computers Can Learn from Us." Good luck and Godspeed.

At least what can be said about the second and third hiatuses of our knowledge—I am reticent to call them gaps due to the natural theological debates they have ignited, sparking as they do ill-fated arguments ending with the unsound conclusion "Therefore God exists," concerning which I have already expressed my opposition—is that Whitehead's thought overcomes them. Instead of inorganic and organic categories, Whitehead replaces them with one, namely organicism, involving relative amounts of spontaneity. We possess a considerable degree of it owing to our free will. This also means, of course, that we can be beset with problems should we avail ourselves of the darker side of our electives. In any case, the second hiatus is bridged through the ability, though limited, of the inorganic world to exercise its own self-determining power, and in this way all of nature becomes organic. As with my criticism of the process system at the end of part 2, I will encourage the reader to decide whether or not this explanation rings true or at least is satisfying.

As for the third hiatus, the problem is alleviated through the exercise of (alphabetically) our aims, desires, dreams, hopes, and purposes. We are all, once again, centers of spontaneity, entailing that we make choices in accordance with our own subjective view of what Aristotle termed final causes—our expectation of our own futures, in answer to the question "Why go to all the trouble and for what?" Essentially, we have a mentality to (also alphabetically) carve, forge, and sculpt out the long-range vision we have for ourselves. Whitehead's deity, further, has such a vision for the entire cosmos and seeks to (once again alphabetically) coax, draw, entice, and lure the world to feel or internalize and grasp that same vision and adopt it as its own. But, contrary to accepting it, we also have the ability to reject God's call and instead call our own shots, since God cannot push or pull but only point to, present, and prompt (the five Ps).

Whitehead's divinity can also potentially cover the first hiatus, in that God containing a world does not imply a world of our current astronomical proportions. At one instant, after all, the universe was a point singularity, about one billionth the size of a proton—not much of a universe, but a universe nonetheless. Then in the interest of room to move, if one is willing for the moment to suspend judgment and entertain conjecture, this drove or gave the impetus for this point singularity to seek and ultimately form the space for it to develop further, its tight quarters not adequately reinforcing the capacity to become.

Admittedly, this is a stretch, but then so might be mathematical models about what we cannot observe, such as the beginning and ending of the universe, for math does not necessarily conform to reality. For example, it is mathematically possible to have a closed, open, flat, or oscillating universe, though not all of them can occur. Compounding this, math cannot tell us why our universe five billion years ago suddenly became an expanding one at an accelerating pace, or if the opposite will occur in the next five billion years, for who knows? We must confess that we do not know how universes behave. Math can afford us models, but they might not get past the speculative stage.

Now the foregoing has more philosophical-theological than scientific backing. Process thought bears explanatory power but not widespread acclaim. Its polling numbers are low, though the amount of support it garners is not an index of its veracity. It may be on to something even if some regard it as quirky or quaint. I have personally no longer thrown my hat into its ring, but it could be a while before we devise a superior metaphysical strategy. It depends on what is more important to those engaged in this quest—bridging hiatuses or uncovering a suitable divinity. Or can we have both?

WHAT'S A BODY TO DO?

Narrowing our focus to the third hiatus, in my appraisal the divisions of humanness are as follows:

Body: that which interacts physically with its immediate environment; and within the body is the blood, which in the Old Testament is considered essential, for life courses through our veins and without it we are corpses: "[T]he life of a creature is in the blood" (Leviticus 17:11; Deuteronomy 12:23).

Brain: a component of the body that receives and processes internal and external sensual and visceral stimuli. Neuroanatomy and neurophysiology do not completely account for the remaining levels.

Mind/mentality: the rational faculty that arranges options on how to respond to these stimuli and reflects on possible outcomes, and is sometimes referred to as our psyche. This is the seat of consciousness and subjectivity/interiority. Many species possess a brain; only a few bear (self-) awareness and (self-) consciousness.

Soul: a faculty that decides upon which of these options to take and with time forms habits—creatures of habit that we are—that mark a person's identity, and is sometimes referred to as our will. That which goes beyond the stages of neuronal firing, upon which materialist descriptions rest, is a response to a state of mind, which in turn is an act of will and constitutes the difference between merely possessing subjectivity/interiority and reflecting on it together with exercising it.

Conscience: a component of the soul which evaluates the appropriateness of these responses and how well they match up to the ideal that a person has set for him/herself and potentially aspires to, and is sometimes referred to as our heart. This is where our final causes, that is, our purposes, are molded.

Spirit: the faculty which enables relatability to one's vision of an ultimate that encompasses these ideals and provides the passion and impetus to reach it. In turn this ultimate sometimes speaks to and urges one's conscience on toward a more refined soul. Were it not for the spirit, we could not function as a living soul, which is the life the Infinite Spirit has given us to be finite spirits and returns to its source at life's end.

Each of these in its own way provides fingerprints which distinguish one person's identity from another, making him or her a unique individual. No two people, not even twins, have the same body, etc. These quadripartite divisions can act in either a bottom-up or top-down fashion; that is, the classification scheme can be read causally upwards or downwards. In living, the body and mind do the physical and mental heavy lifting, but it is not until we reach the soul that we can speak of a person behind the physical and mental apparatus.

The kind of causality as one moves from spirit to body is by way of information—informing the level immediately below it as to the best course of action based on tested and tried actions that have been found to be true or in accordance with the ideal one has adopted. This is information which sparks one into response. There are times, though, when ideals can come under review and be exchanged for what one comes to evaluate as a set more germane to one's new outlook. Thomas Kuhn would refer to this as a paradigm shift; religionists as a conversion.

The soul requires perhaps even a lifetime's worth of value formation and undergoes changes depending on which values it adopts at a specific time. It evolves and constitutes the highest quality that evolution has produced, and it is that aspect of humans which God has the most interest and investment in. The spirit is not the product of natural forces but is provided by the divinity for the life span of an individual. It marks a channel of communication and impression between the heavenly realms and ourselves, either directly from the divinity or via angelic messengers. At death it is removed in accordance with the words of the teacher: "the dust returns to the ground it came from, and the spirit returns to God who gave it" (Ecclesiastes 12:7). The grave is where we sleep and where "there is neither working nor planning nor knowledge nor wisdom" (Ecclesiastes 9:10), though there is an episode to the contrary.

King Saul approached the witch of Endor to disturb, bring up, and consult the shade or spirit of the prophet Samuel for battle strategy in his war with the Philistines, despite warnings against doing so, with mediums and spiritists which he himself expelled (1 Samuel 28). God issued these stern warnings (Leviticus 19:31; 20:6; Deuteronomy 18:11; Isaiah 8:19–20), so there must have been such a prevalent behavior for God to reject. Plus, those in the grave must indeed be recognizable as well as possess some knowledge, otherwise individuals would not have sought them out. The residence of the dead during that time period was known as Sheol, where personal identity survives, but in inhospitable conditions. This underworld was not a place of rest; instead the remnants of a person corroded away.

Nevertheless, both the body and mind experience dissolution at death and the soul is the lone survivor of it. Constituting the makeup of the person, the sum total of all decisions reached is reflected in it. Having a body and mentality in this life enables one to hide or mask or rationalize the true nature of the soul, but it will become evident when the outer layers are stripped away. To God the heart is an open book and no dendrites are required to record habits formed. The refinements of the soul which have not been completed in this life might very well undergo further polishing in the next.

There is a significant aspect of humans which to this point has not been broached in our description. In the debate as to which comes first—whether essence precedes existence as for Plato or existence precedes essence as for Sartre—in the Judeo-Christian tradition we already come with an essence, a foreign one, namely the imago Dei—the image of God. From this perspective Plato is more correct than Sartre, only our essence is not a soul from a realm of Forms but an image from a Person in a realm which has broken into ours by its inauguration through God's Messiah.

Known as the kingdom of God, it is present in part here and now, but its arrival in full is anticipated once the pregnant heaven finally gives birth to a New Jerusalem (Revelation 21:1–2), apparently still in construction, since Jesus mentioned that he is going "to prepare a place for [us]" (John 14:2–3). Its gestation period unknown by everyone except the Father. As a consolation prize, Sartre is right in that we do make ourselves, yet what we make is not our essence but the shape of our souls by way of our choices together with what we have done with and how we have expressed our image in preparation for inspection. Lest Plato gloat over this victory, his view is also faulty since the body is not a prison house for the soul but a five-star temporary accommodation, though even these structures can become dilapidated with time.

It should also be noted that Jesus' saying in Luke 7:28 that "the one who is least in the kingdom of God is greater than [John the Baptizer]" implies that there will be a hierarchy of sorts even in the afterlife, as is the caution that we should opt for the lowest seats at the great banquet in case there be another who is "more distinguished" who will be invited to the greater seat. If not, we might then be invited to "move up" (Luke 14:7–24). Again, there are those who will be called least and great in the kingdom (Matthew 5:19), where we will be refined as though by fire (Zechariah 13:9; Malachi 3:2–3), and with which everyone will be salted (Mark 9:49) in proportion to our deeds (Luke 12:47–48).

From the perspective of the Judeo-Christian tradition, souls eagerly await the reconstitution of their bodies and minds at the resurrection, what God allegedly had planned all along. At this point the quality of our souls will be evaluated for fitness in God's kingdom. Those who have allowed the divinity or its representatives to have access and a shaping influence on their souls will be more prepared for life with the deity in the new heaven and Earth than those who did not. The latter who spent a lifetime rejecting the Spirit's promptings would experience revulsion at the prospect of a protracted period of time in God's company.

They would find it repellent and would seek to remove themselves from it. God will accommodate them as well. Those who persist in retreating from God's entreaties are not compelled to reside in a place where they would rather not be, but a place for them outside is provided for them too. To do otherwise would be cruel. My sincere hope is that the shutting of the doors or gates of the kingdom is not final, but that the opening of eyes might lead to the changing of hearts, in turn leading to an increase in the population of the kingdom. In tune with this, I hope the citizenship office will never close and that more would continue to become naturalized.

FOR WHAT IT'S WORTH

It seems we have no other option than to refer to the above state of affairs as a dualism of sorts. To round out our discussion we will address the following four theological topics: dualism, death and the intermediate state, the resurrection, and the afterlife. The treatment will draw on the biblical statements informing these topics. While we have determined that the Bible has its limitations, and that it resists systematization, for that is a Greek and not a Hebrew innovation, despite its weaknesses it does present a beneficial resource, a useful set of tools to have in one's tool kit.

To start us off, there is an I or self that both has a body and lives in it (Galatians 2:20). The body is termed an "earthly tent," which we can either live in and be at home in or depart from (2 Corinthians 5:1–10). The departing is equivalent to being with Christ (Philippians 1:21–24). The body minus the spirit is a corpse (James 2:26), though while in the body, living according to the spirit is true life (1 Peter 4:6). In relative value, the spirit far outstrips the body, since it is only God's Spirit who "gives life" (John 6:63). Nevertheless, the body is valued as a good creation of God (Genesis 1:31), while even a very good world can still, mythically speaking, contain evil serpents (Genesis 3). And the allotment of time for a human lifespan is variably described as seventy or eighty years (Psalm 90:10) to as many as one hundred and twenty (Genesis 6:3).

Regardless of their longevity, bodies will eventually be "put aside" (2 Peter 1:13–14). The apostle Paul was even given a vision of the heavenly realms, which he suspects might have been an out-of-body experience, despite his not divulging that it was he himself who had it but kept it unconvincingly hypothetical (2 Corinthians 12:2–3). Unfortunately for us he was not permitted to develop this theme but was served with a gag order (verse 4). In any case, this putting aside is called having "fallen asleep" (1 Corinthians 15:18; 1 Thessalonians 4:13), and is theologically known as the intermediate state after death. Yet "we will not all sleep" (1 Corinthians 15:51), for bodies which are sown, that is, interred naturally like seeds, are raised spiritually and are no longer perishable (verses 42–44). Most importantly, despite the body being destroyed, God has prepared for us a "heavenly dwelling," building, or house, and "to be away from the body" is to be "at home with the Lord" (2 Corinthians 5:2, 7). And should we have lived by the Spirit, our bodies "may be destroyed" but our "spirit[s] saved" (1 Corinthians 5:5), for we need not fear "those who kill the body," since only God "can destroy both body and soul" (Matthew 10:28).

Aside from sleeping, there is one New Testament passage suggesting that at least some in the intermediate state are conscious. First Peter 3:18–20 postulates that the Spirit of the deceased Jesus was transported (contrary to the Apostle's Creed, there is no textual confirmation that the journey amounted to a descent, only the traditional view that the direction must be downward) to those in holding cells where Jesus made an appeal for them to alter their allegiances. Apparently they had been held there ever since they disobeyed at the time of the mythical Noah, thereby reassuringly informing us that the opportunity to walk with the Messiah has not as yet run out and might not end with physical death.

What is more, if the spirits of Moses and Elijah really did speak with Jesus on the occasion of the transfiguration (Matthew 17:1–8; Mark 9:2–8; Luke 9:28–36), then there is at least some of both knowledge and consciousness in the intermediate state. Now of course all this assumes, as it does with any isolated scriptural passage, that each of the writings can be taken as bearing equal weight with that of any other, but this is not automatic. More work must be done than simply stating one's personal preferences if it overlooks or ignores other passages that could speak to the same issue. To do otherwise is negligent at best, propaganda at worst.

The notion of raised or awakened bodies is not confined to the New Testament but is also attested to in the Old in both early (Isaiah 26:19) and late (Daniel 12:1–2) writings, though it is not until the New when we are given a glimpse of it. One interesting note is that Jesus' resurrected body was demonstrated only to a chosen few (Acts 10:40–41), implying that it was not an objective event, and bore enviable powers such as transporting through locked doors (John 20:26), showing that it was not limited by materiality.

Additionally, it was possible not to be cognizant that one was in the presence of the new version of the old Jesus, meaning "they were kept from recognizing him" (Luke 24:13–16), and that some thought they were seeing a ghost (verse 37), the first instance of which occurred when the earthly Jesus approached the disciples in their boat while he was walking on the Sea of Galilee toward them (Matthew 14:26; Mark 6:49). The unrecognizability occurred not only from a distance of approximately the length of a football field but also up close and personal, his disciples having clued in only through his demeanor (John 21:4, 8, 12–13). Perhaps most significantly, it was possible even to doubt the event itself despite standing in front of the risen Jesus and even among those same chosen few (Matthew 28:17; Luke 24:38).

Paul seems to think that there is a major difference between our current physical state and our resurrected one, so much so that the two are beyond comparison. In the gospels, however, all of which are writings later than those of Paul, Luke in particular in the Emmaus road passage has the risen Jesus accompanying two others who failed to recognize him. For the present issue at hand, the concern is not so much the recognizability of someone's risen identity but the fact, at least according to the accounts, that the individual appeared as an ordinary person with no extraordinary features. Jesus' interlocutors on the Emmaus road were not taken aback as the disciples were when the pre-risen Jesus came to them in their boat while walking on the water. In the latter instance, they were so perplexed that they thought they were seeing a ghost, whereas it was the usual Jesus they had come to know.

Much different is the Emmaus road encounter, where the risen Jesus did not resemble anything alien in the least but appeared as a regular individual. How can this be if Paul understands the difference to be so vast? Are the gospels to be taken as the more mature rendition, if only for the simple reason that they came later, or are the writings in closer proximity to the events in question more reliable than the more recent ones? Which version can we have more confidence in, or can resurrection bodies shape-shift so as to assume a more customary form? And if resurrection bodies are to be similar in type to angelic ones, then how can Jesus' resurrection body be so, well, standard human issue? Ambiguity prevails.

While in the intermediate state, souls are separated from bodies, and those who have been martyred are ushered before the heavenly court or under its altar (Revelation 6:9; 20:4). John of Patmos was given this vision, meaning souls can be seen, although it begs the question as to how a mere human can recognize them. Subsequent to souls being reunited with their reconstituted bodies, the Day of the Lord will involve a reckoning of the deeds "done while in the body" (2 Corinthians 5:10), and regrettably, as intimated, some will find the door to the kingdom shut (Matthew 25:1–13; 2 Thessalonians 1:9). The nature of a settled past is such that it cannot be edited as an author edits a manuscript.

Other aspects of the hereafter include the following. Initially, time will continue to have a meaning in the next life. For instance, there will still be a concept of before and after, as John states that "After this I looked . . . " and "After this I saw . . ." (Revelation 4:1; 7:1). This even extends to our normal way of marking chronology: "there was silence in heaven for about half an hour" (8:1). Next, there seems to be a discrepancy between two passages concerning food. On one side, Paul in Romans 14:17 announces that "the kingdom of God is not a matter of eating and drinking," over against

Jesus' parable of the wedding banquet (Matthew 22:1–14). Compounding this is the notification of the reinstating of the tree of life, which will yield a monthly harvest, as well as the river of the water of life (Revelation 22:1–2, 17), indicating that heaven will have both temporal and digestive elements. After all, the risen Jesus was said to have had a taste for seafood (Luke 24:41–43).

Another of Jesus' parables also implies that the afterlife will be active instead of passive when he declares, "I will put you in charge of many things" (Matthew 25:21). At the very least there appear to be administrative vocations to discharge (likely appointed rather than posted competitions). Lastly, for those who are encouraged by and trust in a universalism in the hopes of being reunited with friends and family on the other side, there is this comforting statement: "God, who is the Savior of all [people], and especially of those who believe" in the gospel (1 Timothy 4:10). Note that the text reads "all" and not "only believers." And to top it all off, God's followers will find themselves bestowed with "life forevermore" and "immortality" (Psalm 133:3; Proverbs 12:28), in God's company, where the video portion is resplendent and the audio captivating, all to take place in the New Jerusalem, where, thankfully, the city "gates [will never] be shut" (Revelation 21:25).

APPROACHED BY AN ANGEL

It also seems appropriate at this point to submit an installment regarding angelic beings. To begin with, they are given one personal name, two of the most important being Gabriel, who interprets visions (Daniel 8:16-26; 9:20-27) and heralds the birth of the Messiah (Luke 1:11-20; 26-38), and Michael, who protects Israel (Daniel 10:13, 21; 12:1) and bears a rank, namely archangel (Jude 9). (Are we to assume, then, either that by having but one name there are fewer of them than there are of us, or that God is more imaginative when it comes to naming? We ourselves also possess some such capacity, given that the mythical Adam was mandated to name all the animals [Genesis 2:19-20]).

Perhaps this rank is one of military superiority, for "there was war in heaven. Michael and his angels fought against the dragon, and the dragon and his angels fought back" (Revelation 12:7), evidence that God is not unfamiliar with battle campaigns even at home. Despite such a grand campaign set out for them, they have more mundane tasks as well, since they are spirits sent by God, "mighty ones who do God's bidding" (Psalm 103:20), to minister to and serve humans (Hebrews 1:14) and guard them (Psalm 90:11; 103:20), particularly waiting on Jesus after his temptation (Matthew 4:11; Mark 1:13). Other verbs employed to describe their role include lifting up (Psalm 90:12), driving out opposition (Exodus 33:2), encamping around, delivering (Psalm 34:7), and saving God's people (Isaiah 63:9).

Additionally, they were commissioned to "put the law into effect ... by a mediator" (Galatians 3:9), and when the mythical Abraham received three visitors, seemingly human but actually angels of the Lord (Genesis 19:1), they sampled food prepared by Sarah (Genesis 18:8). One appeared to Moses in a burning bush (Exodus 3:2), and during the forty years of wilderness wandering the Israelites ate quail and manna, the latter referred to as "the bread of angels" made from "the grain of heaven" (Psalm 78:24-25) (prepared by a superior baker and likely gluten free). Hence they have the wherewithal to ingest, but do they also digest in common with humans? And this is not all. Both the Abraham episode and the short story above demonstrate that occasionally they take on human form and we unknowingly interact with them (Hebrews 13:2), though Abraham was fully aware.

Should this be the case, certain logistical issues surface, such as who their barbers and tailors are and which trademarks can be found on their attire.

They further appear to humans in dreams and visions (Ezekiel; Matthew 1:20), deliver revelations (Revelation 1:1), and they lack omniscience but have inquiring minds, in that they would like to know more about the hidden things and mysteries of the gospel (1 Peter 1:12), potentially implying that we know something they do not, or that they and we are equally unawares. Michael has also been known to enter into heated "disput[ation] with the devil about the body of Moses" (Jude 9). While on the topic of the demonic, some angels can and also have fallen from grace in not having kept "their positions of authority but abandoned their own home" (Jude 6). Furthermore, the devil demonstrates knowledge of the Old Testament by citing it when tempting Jesus (Luke 4:9–11).

And not least, back to the topic of angels, they are called upon to weed out evil (Matthew 13:41). Essentially, we have been made a little lower than they are (Psalm 8:5; Hebrews 2:7), as was Jesus himself (Hebrews 2:9), though we will become like them in reference to the next life as not involving marriage (Matthew 22:30; Mark 12:25; Luke 20:35–36), perhaps entailing that they now and we then will be without gender (and, as some might disparage, quite possibly hormones), even though the church is referred to as the bride of Christ (Revelation 19:7–8; 21:2, 9). Finally and highly significantly, we will be given authority to judge them (1 Corinthians 6:3).

According to popular assumption, if every human is assigned one—a guardian angel of sorts, then there are at least as many of them as there are of us at any given time, unless of course they do double duty or more. Whether their number is as great as the amount of persons ever having lived in human history, or whether angels have subsequent assignments once a given individual passes on, is another matter. What the Scriptures do declare is that there are at least "thousands upon thousands" of them (Hebrews 12:22; Jude 14; Revelation 5:11), even "ten thousand times ten thousand," meaning a great throng (Revelation 5:11). Compare this to how many persons the Scripture promises there to be in the end: "a great multitude that no one could count from every nation, tribe, people and language" (Revelation 7:9).

The terms employed above paint angels as what has been referred to as the "causal joint" between God and the world, thereby enabling God to interact with the ranks of the world through these higher-ranking officers. As creatures they possess bodies of some description and straddle the spiritual and material, holding one foot in this world and one in the other, and as a result allowing the ethereal angelic rubber to meet the substantial earthly road as a type of hybrid operating in both worlds as divine go-betweens. After all, when we observe divine activity in the Scriptures it is often carried out by the angel of God as God's representative (Genesis 19:1; 22:10, 15; 32:1; Exodus 3:2; 14:9; 23:20–23, as examples). Hence they could be called the executive branch of the divine administration.

Anticipating an objection, some might be inclined to assert that this view, contrary to Ockham's razor, multiplies entities unnecessarily—first God and then angels. As a counterpoint, at least from the standpoint of the sacred text we are examining, angels do seem to be part of the package. In a similar vein, all of the foregoing, though debatable, is in terms of working with the biblical statements as bearing equal weight, aside from issues of historicity and reliability. The exercise is geared toward reflecting on the statements as they appear at face value (prima facie) for better or for worse. Criticism may be leveled aplenty, sprinkled with a generous helping of skepticism, yet it is important to establish how the Scriptures read as they stand for clarification purposes.

YOU AGAIN!

The idea of reincarnation is neither new nor exclusively Eastern, though its origin is. It stems from two great world religions: Hinduism, where the concept is known as the transmigration of souls; and Buddhism, where the notion is termed codependent origination. This line of inquiry made its way westward to Persia, where it may very well have been adopted by Zarathustra (Zoroaster). It was there that the Jews were potentially exposed to the doctrine while in exile during the Babylonian captivity. Egypt also figures into the calculations. The journey west continued where it found its way into the West proper, specifically in the ancient Greek world. It entered the thinking of at least two pre-Socratic philosophers, namely, Pythagoras and then Empedocles. Following this, the position entered Socratic-Platonic reasoning from Plato's Myth of Er in *The Republic* and three other dialogues.

In Plato's outlook, souls fall into bodily incarnation in a cycle of rebirths owing to past life misdeeds. Souls descend, not of their own volition, to be incarnated into bodies of their own choosing in a limited selection process cast by lots. Hence the soul which finds itself in an in-between state, having inhabited a physical human form that has passed away and is awaiting dispatch into another human, has some choice in the matter. This is different from Eastern accounts, where karma enters in and becomes the deciding factor as to where the next habitation will be, which could even reside in an infra-human species should the prior life have been less than commendable or even adequate in terms of moral rectitude. It might be getting its just desserts. In either instance, Plato in the West or schools of Hinduism and Buddhism in the East, there is no cognizance of what transpired in the former life, unless this can be teased out by certain contemporary retrogression hypnotherapists. For Plato, or Socrates to the extent that Plato is his teacher's mouthpiece, the awaiting soul takes a dip in the waters of the River Lethe—the river of forgetfulness—where it loses any recall of previous life events. Interesting that water should be selected to perform the feat, since, touted as it is a universal solvent, it effectively dissolves even the soul's memory of its past.

One of the criticisms of reincarnation is the contention that we cannot learn from past lives so as to correct our behavior if we do not possess knowledge of these lives from which to work. Plato was ready with a response to meet this challenge: knowledge obtained in a former life is easily retrieved in the next, since in education we are essentially recalling what we already knew. According to Sheldrakean reasoning, as an aside, he might quip that perhaps this is why we do not readily pick up Latin these days. The difference between ancient Jewish and Greek understandings of the viewpoint is that in the former the spirit is taken as the feature which migrates from life to life, while for the Greeks it is the soul. The Western world opted for the Greek version.

Reincarnation moved into the Common Era through several church fathers, the Neoplatonist philosopher Plotinus, and later through a host of stellar literary and philosophical figures, particularly German, French, Russian, as well as British and American authors. While these persons may not be widely seen as champions of the position, they do have their reincarnation moments in the subject matter of their written works. The church took its stand on the issue by the Fifth Ecumenical Council, known as the Second Council of Constantinople, in 553, where the doctrine was condemned and those holding it were anathematized (excommunicated).

A possible Christian stance on reincarnation could be the assertion that God wishes to discern whether or not the same soul will make the same mistakes in multiple lifetimes, or to give it yet another opportunity to get it right. The difficulty for orthodoxy (classical theism) here is that the view would have negative implications for God's alleged omniscience of the future, for a divinity which knows the future with certainty would not need to experiment with souls in successive lifetimes so as to determine what a soul would or would not do the next time around. For those with no such investment, however, the problem is not so dire.

There is a tendency on the part of the two sides of the debate to adduce what they consider the most telling scriptural passages which they believe speak to the issue. There are ordinarily assumed to be two main ones: for the anti-reincarnation camp, they argue that humans are "appointed" (KJV) or "destined to die once" (Hebrews 9:27); and for the pro-reincarnation contingent, John 9:2–3 suggests just the opposite—"[Jesus'] disciples asked him, 'Rabbi, who sinned, this man or his parents, that he was born blind?'" The idea was not foreign to the disciples, who, when asked by Jesus as to who it is that people say Jesus is, answered that his identity was one of the prophets of the recent (John the Baptizer) or remote (Elijah) past who has arisen, that is, come back to life (Matthew 16:13–14; Mark 8:27–28; Luke 9:18–19), though this could also be interpreted in a resurrection-oriented fashion. In any event, each position has its fragilities.

In the first case, dying once is not without its biblical exceptions, namely Enoch (Genesis 5:24) and Elijah (2 Kings 2:11–12), who were allegedly whisked away while still very much alive, and Lazarus (John 12:9–11), who had the unenviable fate—unless one is part of the pro-reincarnation crowd—of dying twice, possibly violently for the second. Plus, the Enoch example as definitive is not patently plain from the wording: he "walked with God; then he was no more, because God took him away," for being no more and taken away could even be interpreted as simply having passed away.

In the second instance, the reckoning is that only the blind man himself could have sinned were there to have been a previous life for him in which to do so. Yet the conventional interpretation of this passage involves Jewish purity laws, for if a pregnant woman were to be unclean or impure, then all parts of her would be deemed as such, including the contents of her womb. Hence this passage need not be interpreted in a reincarnation manner, and thus neither case exemplifies a clear advantage over the other. This leaves us, once again, with an ambiguity, since the biblical evidence could be understood two different ways. Whatever we think about the issue, therefore, the Bible will not assist us much, as it does not present one monolithic stance on these and other concerns.

The assumption on the part of some and the hope on the part of others once again is that God offers humans every opportunity to mend their ways. We cannot be certain as to how far the orthodox portrayal of God's grace and mercy extends, but if it does beyond this lifetime then there are some caveats or rules of the game to abide by. If, for example, there is no purgatory as such, or analogously as hockey players are intimately familiar with, specifically the penalty box, where albeit short sentences are served and once completed those who find themselves in it can resume their participation in play; and if Irenaeus is incorrect in that there is no two-step approach where in this life we are made in the image of God and in the next we are fashioned in God's likeness where the transformation process continues after death; and if there is not even any type of afterlife refinement stage to undergo—and that is a lot of ifs—then all that remains is multiple lifetimes. This is not a knock-down argument, nor is it required to be. All that needs to be established is that reincarnation cannot be ruled out from the outset as a way to improve our condition beyond a single go-round, and at least this much can be regarded as fairly firm.

Questions which persist, in any case, include what motivation there might be for individuals to get it right this time around if one has additional lifetimes to work with. The stock response often is, "What is undesirable is to descend the great chain of being next time around into an infra-human species or even a plant," to which a rejoinder could be made that in the grand cosmic scheme of things this might not be so bad if one eventually ascends back up the scale once again. And on the topic of schemes arises another question, particularly who or what is in charge directing the soulish traffic? If the standard comeback takes the shape of a karmic proposal, then this begins to sound conspicuously like a conscious agency with a design in mind, reminiscent of a divinity, which is unlikely what at least some reincarnationists would intend.

CAN THERE BE ESP?

If radio, films, and television shows in recent history are any indication of what a sizable proportion of the listening and viewing public takes to be themes both timely and topical, then the paranormal has definitely captured the public's imagination. Programs stretching from *Coast to Coast AM* to *UnXplained* suggest a significant interest on the part of many people, even well-educated ones, to delve into the unusual (to say the least). Despite the threat of brow-beating from some practitioners within the scientific community, the interest in things parapsychological endures.

I suggest the term paraordinary despite its clumsiness in place of paranormal so as to avoid the noticeable value judgment. Parapsychology, however, is a useful term since it refers to possible aspects of the mind not customarily attributed to individuals or investigated by the discipline of psychology. Besides, little of this kind of research is widely known simply because it is often considered peripheral to knowledge at best or pseudo-scientific at worst. Perhaps a course of study is warranted to help organize our thoughts and determine what might be worthy of further scholastic pursuit. Ideally, an academic setting would address these subjects in a more rigorous fashion than that which can be obtained through the popular press. Yet given the conservative nature of academic institutions, incorrectly understood as avant-garde, they are self-appointed gatekeepers mostly given to upholding the status quo of acceptable subject matter, so I am not holding my breath.

There are a few institutes of higher learning, though, which have already undertaken such a quest, and for them the type of courses I have in mind are either currently being taught or could benefit from them. They could inform us not about what to think but how to think in reference to these issues, so that in our scrutiny of them we may distinguish between the credible and the incredible, the knowable and the unknowable, and whether this form of ignorance is in principle or merely in practice.

The proposed course could be entitled "Boundary Abilities," covering those experiences on the borderland between physics and metaphysics. The attempt would be to disclose the extent to which certain phenomena should be regarded as outside the confines of science and maybe even illusory, or as legitimate albeit mysterious scientific anomalies. An additional objective of the course would be to uncover whether these phenomena are more properly the province of religious encounters or metaphysical speculation. But what are some of these phenomena we are referring to?

For starters, some definitions are in order. Clairvoyance, from the French, means seeing clearly and refers to knowledge of objects and events in the world without access to them. Some might be hidden from view, others removed in space or time. This may involve not only lost objects but their history and whether they were used in, say, a criminal act. Telepathy means awareness of the thoughts of another person. Both of these abilities fall under the category of extrasensory perception (ESP). Precognition affords the bearer to have an indication of occurrences prior to their onset, and telekinesis permits one the capacity to move an object through thought power alone. Other powers include remote viewing, channeling, and astral projection. We will make mention of the second and third listed here.

Whitehead admitted to telepathy as a distinct possibility, though he never developed this line of inquiry in any of his extant writings. He has a few obscure references to it in his major works, but nothing more (who knows, however, if his unpublished notes could suggest otherwise). Despite my reservations about his system, Whitehead, like the Bible, has not outlived his usefulness and, also like the Bible, retains pearls of wisdom.

But does telepathy cohere with his system? For him, all that experiencers can know is the data in their own environments in both the immediate and distant past. Knowledge of neither the present nor the future is available to us, with God as the lone exception having access to contemporaries. Still, only the past can yield knowledge for us. Our experience of our own interiority, such as memories, is what we know most intimately—no one else has immediate access as we do to our own subjectivity. Thus knowledge for us can come from within as well as without. Knowledge is not confined to the experiences contained in our own selves, but we can know other experiences since we are internally related to them—we feel or prehend them in Whitehead's terminology.

As subjects, we have an attitude about the data contained in the objects we are presented with in our past. This was true when those objects were themselves subjects, and will be true when subjects become objects and are felt or experienced by newly forming subjects. How a subject felt about an object is recorded when this subject becomes another object. This feeling can then be perceived by a new subject. This is not necessarily through coming in contact or colliding with that object, for that would be the language of the conventional substance view of science. But is this the language of the telepathic?

Whitehead informs us that existence is exhaustively composed of momentary occasions of experience, events which receive input from the objectively settled past (as being) to a subjectively relative present (as becoming) and here respond to them. Given that for the present author not only am I all about religion but also sports, let's take an analogy from football. Let's say a quarterback got sacked on a previous play. Chances are s/he objects to this occurrence and may even be deeply offended that it happened to him or her. S/he has immediate access to the event, after the fact, through memory. I can well imagine that s/he would not want to suffer the same indignity again.

S/he is internally related to the experience of the quarterback who just got sacked on the previous play, for it was him or her. The experiential access to this event Whitehead called prehension. Sometimes spectators or a television audience will grimace when such a hit takes place. They might even have vicariously felt it themselves in their own experience. They may know what it is like and wince, "That's gotta hurt," or "That pain is gonna linger," or "That's gonna leave a mark." This is analogous to what is intended by both Whitehead and telepathy—one internalizes what another has felt and thought, respectively.

Now every analogy ultimately breaks down and this one is no exception. To amplify, some may be inclined to think that the process perspective does indeed have a place for telepathy as it occurs each time an entity prehends, but this line of reasoning encounters a drawback. If the assumption is that internal relatedness is equivalent to telepathy, regrettably this is wide of the mark. Telepathy is the transaction of communication of thoughts at the level of mind to mind, and when received it derives from the past, whereas prehension is about the transaction of information of feeling at the visceral level, also from the past, which does not strictly amount to telepathy. And while empathy is not the same thing as telepathy, the latter can be considered an implication of Whitehead's scheme; its route would just be indirect, the cognitive by way of the visceral.

Yet parapsychology is also the point where his view is limited. Since perception for us is entirely of the past, it rules out anything like precognition. Future events are completely closed off from view in his strategy, even for God, who is the lone entity which enjoys access to the contemporary world. Alleged precognition must then somehow be forced into a past event if we are to retain Whitehead's categories as they stand. But if someone actually enjoys access to or can tap into the future in a way that reduces its contingency, making it more or less a foregone conclusion, then Whitehead's model would require serious reformulation.

Bookies who give odds on sporting events, allowing others to wager on their outcomes, would certainly like to tap into such a gift, talent, or sensitivity. They would likely be the first in line to receive some were it true. But for those who are keeping score, the Jesus of the Scriptures is touted as having this ability. He either has or was given telepathic knowledge of the thoughts of the hearts of his hearers (Mark 2:8), he knew what others were thinking to themselves (Luke 5:21–22; 6:7–8; 9:47), and knew all persons and what was in them (John 2:23–25). For him nothing seems to be private. Interestingly, the gospels state his alleged ability matter-of-factly, without anyone exclaiming, "How can you know that?" even if they were taken aback. Can others also possess such a capacity or, once again, is he the lone exception?

The other major thinker we are considering here also weighs in on the topic of ESP, specifically telepathy. Sheldrake chimes in with the contention that minds are not confined to brains but extend beyond them in both time and space. Minds further survive the death of the body, making Sheldrake a dualist of a certain stripe. He reasons that eyes and minds do more than simply receive information from outside, as if to say that one's vision of an object, for instance, is all in the mind and has no reality above registering in the brain. For him, vision additionally projects outward so as to mentally make contact with the objects of experience. This phenomenon is felt most noticeably when individuals have "a sense of being stared at." Minds project externally as though touching another person, prompting the latter to turn around in order to determine whether someone is looking. A definite advantage for survival purposes.

Abilities of this sort extend into the telepathic in examples such as thinking of a person followed by having that same person call on the phone, as well as mothers having an awareness of their infants requiring their assistance even when they are not in earshot. These capacities are adaptive and have an evident selective advantage despite becoming less prominent in the electronic age, though even now have not been muted. Moreover, such sensitivities have also been found in the animal world, where pet dogs, among others, have demonstrated an awareness of "when their owners are coming home."

Sheldrake has conducted experiments on all of these phenomena and his results are striking, but he may be overreaching when claiming that some animals seem able to predict upcoming disasters such as earthquakes. What might explain their anticipatory behavior more accurately is sensing certain vibrations, scents, or other signals unnoticed by humans, emitted, say, by the Earth as precursors to these events. Yet even if so, this would not subtract from the significance of their ability.

Anthropologists inform us that the human brain in its evolutionary history increased in size, but at a certain point, shall we say, overshot the mark and became reduced by ten percent, whereupon the size plateaued and stabilized, though might now be experiencing an increase once again according to recent investigation. Were there any previous abilities that were either diminished or lost by this reduction? Did we possess greater sensitivities than we currently do? Was greater ESP one of them? One would imagine that bearing them would have been adaptive, advantageous and selected for. Or, theologically speaking, is this what the myth of the Tower of Babel (Genesis 11) was driving at, specifically God perhaps confounding our mismanagement of these or other capacities? One more thing, the belief that all behaviors extant today are exclusively the products of natural selection is false, since there are behaviors which remain selectively neutral, like coyotes howling at the moon. It is not a mating ritual, for example, and this type of reasoning should caution us about what we consider selective.

I would be remiss if I did not at least also mention what some describe variously as a trend, fad, hype, or hysteria about UFOs. One researcher who has conducted a study of the phenomenon came to the conclusion that, in addition to the plasticity of the brain's dendrites, there is also a plasticity of memories as already alluded to above. We become conditioned to expect the world to behave in the manner to which we have grown accustomed, and when the new is encountered we attempt to make sense of it in terms of the categories we have unquestioningly assumed and become comfortable with. Much like what Thomas Kuhn alerted us to about paradigms that have a hold on us, it takes a crisis before we lose confidence in them and would much rather retain our domesticated worldview at any and all costs. But there comes a time when we need to admit that it must yield to a new and improved model (which might actually be an old and refurbished one). Our memories meet up with our expectations as well.

As Susan A. Clancy (while at Harvard!) observes, "our brains engage in an act of construction when we call up our memories,"[12] and this reconstruction occurs with the passage of time such that "we make up history as we go along."[13] Our memories become modified and framed by our expectations and this can be applied to the alleged UFO alien abduction accounts.[14] Clancy diagnoses these cases as "long[ing] for contact with the divine, and

---

12. Clancy, *Abducted*, 68.
13. Clancy, *Abducted*, 69.
14. Clancy, *Abducted*, 142.

aliens are a way of coming to terms with the conflict between science and religion. I agree with Jung: extraterrestrials are technological angels."[15]

She then proceeds to make the startling claim that what "people get from their abduction beliefs [is] the same things that millions of people the world over derive from their religions: meaning, reassurance, mystical revelation, spirituality, transformation."[16] She compares this kind of experience to the beneficial features certain religionists obtain from their ritual and prayer: "they're happier, healthier, and more optimistic about their lives than people who lack such beliefs. We live in an age when science and technology prevail and traditional religions are under fire. Doesn't it make sense to wrap our angels and gods in space suits and repackage them as aliens?"[17] She wishes she had some of her own.[18]

My comments are these: not everyone benefits from alleged abduction experiences and the trauma which some undergo can become psychologically crippling. I wonder whether her sample is sufficiently representative of the abductee population. Differences between religious experiencers and abductees include, for some, the latter being unsought-after and not the kind of thing one would wish on another. Perhaps specific persons already have a fulfilling purpose to and meaning in life and are not seeking any more, and for whom an abduction experience would constitute an unwanted intrusion. Plus, I question if there would be many who would proselytize others to seek abduction experiences of their own outright for the benefits they could bestow.

For a final look at parapsychological themes to be addressed here, and since it has potential bearing on the nature of the afterlife, are instances of near-death experiences (NDEs). They frequently involve out-of-body experiences (OBEs), where hospital patients, usually during surgery or attempts to revive them, have the impression of floating above their bodies and looking down upon them from above as well as efforts of staff to treat them. Reports from those who have encountered NDEs and/or OBEs often include a vision of a bright light at the end of a dark tunnel.

This type of experience, however, need not be out of the ordinary but can have a natural explanation. It occurs when oxygen and blood supplies are diminished to the extremities, specifically the eyes and brain. I have encountered a similar depleted condition myself along with the tunnel phenomenon when having engaged too vigorously in cardiovascular exercise

---

15. Clancy, *Abducted*, 154.
16. Clancy, *Abducted*, 154.
17. Clancy, *Abducted*, 154.
18. Clancy, *Abducted*, 154.

when having eaten insufficiently beforehand. Consequently, the least that can be said about such testimony is that not all paraordinary events are without ordinary elements. But on the contrary, it is difficult to discount the new knowledge that OBEs can provide whereas through ordinary means it would elude observation. Reports of sights and sounds while under general anaesthesia are notable, though they could be obtained via telepathic means.

I suppose interjecting my own personal experiences would be warranted at this point. They are not plentiful, nor do I seek them out. What few of them there are, instead, have sought me out. For instance, I am the recipient of vivid, sometimes lucid dreams, mostly during the springtime for some unknown reason. Two of them have been embedded in my memory even without the aid of recording the particulars immediately subsequent to awakening. There are other dreams I recall, but purely on the basis of my having jotted them down upon arising; otherwise by now they would have been lost to the mists of time. Yet those two have persevered with me in clear detail ever since they occurred. I purposely did not write "ever since I dreamt them" because I am still uncertain as to the origin of dreams in general and if we must be their authors. In short, do they arise from within or without? And are they a gift or an affliction?

In any case, the latter in time of the two dreams is inconsequential. It involved a coelacanth, a marine animal long-thought to have gone extinct, but a specimen of which was found off the coast of east Africa in 1938. In the dream, one had found its way into my parents' domicile and my mother had the ill-advised idea of baking it in the oven. Like I said, inconsequential, and I fail to comprehend why this specific dream should have stood out in my memory. The former dream, on the contrary, is too unsettling to relate; I do not even care to divulge its contents. I have no difficulty recalling this dream, perhaps because it is too unnerving and private to forget.

There is one event, however, that my wife and I shared of which I do feel at liberty to render an account even though it rattled us at the time. It took place during the time she and I lived in a Canadian city next to Niagara Falls named St. Catharines. One afternoon I was in the basement of our bungalow, in a room where we kept all our stereo equipment, when I noticed Alice entering a room at the opposite end of the basement. I called out to her but no response. Reckoning that she had not heard me, I went over to the room myself. She was nowhere to be found and there was nowhere to hide. I left the room puzzled when suddenly she came down the stairs to where I was standing. She had a dumbfounded look on her face and announced that she had just seen me in the upstairs living room walking away from her. I then told her my experience. We looked at each other in bewilderment, wondering what feat of nature could have generated this phenomenon—each seeing an image of the other simultaneously. An explanation for it has eluded us to this day.

Spectres of those emotionally close to persons who have just passed away unbeknownst to these friends or relatives have been reported as having appeared to them and later determined to occur at the moment of their passing. But both of us are still here. Could we really have been imagining each other's physical form manifesting at the very same moment? We were not even thinking about each other at the time. As with my dream question, I must ask whether these visions surface from within or without. I am not content merely to relegate these experiences to the file drawers or dust bins marked as anomalies, curiosities, oddities, or cold cases, but I am at the same time not entirely sure as to how to proceed, nor are the prospects patently plain.

SOMETHING CANNOT ARISE FROM NOTHING

Alluding to aliens of an extraterrestrial variety, the following historical tale comes to mind. Cosmologist Fred Hoyle's steady-state model of the universe was the major theoretical competitor to the big bang theory. Note the past tense. In an attempt to distinguish his view from the other, he even went so far, in a BBC radio interview broadcast in 1948, as to name his opponent's view: the big bang, and it stuck. Hoyle teamed up with colleagues Thomas Gold and Hermann Bondi to augment Albert Einstein's insistence that we reside in a static universe, where there is no net change with time.

Hoyle and associates (not a legal firm) contended that the universe remains the same throughout time in terms of overall temperature and density, and the way this is accomplished, they suggested, is simply the introduction of a single atom of hydrogen in every cubic centimeter of space at a rate of one every million years. They believed this to be the sufficient condition for Einstein's static universe to hold. The trouble, though, is where this added alien hydrogen originates. They had an answer for that. They claimed—I hope you are sitting down—that it came from nowhere! Really? Yup, they were serious. How this is an improvement, one must despair, over the traditional Christian doctrine of creation *ex nihilo*, or creation out of nothing, that they sought to supplant is not immediately clear. One might be inclined to declare that the universe is bereft of this capacity.

The impetus for their program was to oust the notion of a universal beginning, for the understanding at the time was that traditionalists would point to a beginning of the universe as the point where the finger of God could be posited. The counter-strategy was to remove this need, since erasing a beginning would entail the removal of any corresponding God requirement. This is an example of the drive to undermine a theological doctrine and it became the motivation for their science, but a non-scientific interest should not be the deciding factor for a scientific program.

In the spirit of discovery, this was not science but propaganda, for they intended to overthrow a religious position by imagining an astrophysical situation that would accomplish it. And all, it must be emphasized, without evidence. No new hydrogen is putting in an appearance anywhere in the cosmos, and lack of observation is normally a clincher, despite the old chestnut about absence of evidence as not equivalent to evidence of absence. There has been sufficient space-time for such an event to occur, however, such that this hydrogen is now without excuse. Also, while true that particles in space go in and out of existence incessantly for nanoseconds, the point is that they go back *out* of existence; they do not remain as the steady-state theory insists they must.

What Hoyle and his team had in their crosshairs was a particular dogma, and with this as their focus they betrayed their own religious posture. Moral of the story: just stick to doing actual science and leave any religious implications to watercooler conversation. And should one elect to enter the debate oneself, then a lab coat will not afford one any philosophical advantage, nor does it give one's own ideological views any more credence. Oddly enough, Hoyle later changed his views and became impressed with the complex folding of biomacromolecules spoken of earlier in reference to Sheldrake and crossed the floor to join the Intelligent Design party.

Molecules like hemoglobin, it will be recalled, fold into a complex three-dimensional pattern in only one way among several possible ones—the elected pathway being monotonously singular. As such, Hoyle dispensed with his earlier atheistic stance and the later Hoyle became an ID enthusiast (the present author making no value judgment as to whether this constitutes an improvement, for advancements, like the bigger, are not always better. Yet even though I decline the invitation to weigh in on the issue, having already done so elsewhere, out of honesty I make the lone declaration that I throw my hat in the ring with its detractors).

By way of explanation, the main difference between evolution and ID is that proponents of the latter believe that certain organismic and cellular structures, such as the eye and microtubules, are so complex and sophisticated that they could not have arisen via stages as in evolutionary theory, but must have appeared on the scene all at once through divine assistance, whereas evolution takes these steps as actually having occurred and when they did so it was through natural forces alone. For instance, an IDer could be heard asking, "What are the stages of a whale's blowhole and how is it adaptive and advantageous along the way so that they would be selected for by natural selection even while incomplete?" Eventually, the evolutionary approach has consistently come to light as being the correct one in both these and other instances.

Once again Hoyle has fallen victim to the trap of taking scientific findings to religious conclusions, an example of the strategy known as natural theology, a view which I have already outlined and critiqued. And curiously, contrary to the prospect of a beginning to the universe suggesting a divinity—a proposal that can be traced back to the Belgian priest and astrophysicist Georges Lemaitre in 1931—the tendency in more recent times is that a beginning to the universe, whether through the big bang or something else, is religiously ambiguous, for by itself it neither points conclusively toward nor detracts from the possibility of the existence and activity of a divinity.

In any event, the steady-state theory has been abandoned and Einstein's static universe discarded. (Einstein did introduce a cosmological constant as a type of buffer or fudge factor to hold the universe in check, but later considered this to be the greatest theoretical blunder of his career. His blunder, though, was our dividend, for it later became the key to understanding the very movement of our universe, just not in the way he anticipated.) In fact, the universe is currently experiencing a runaway expansion rate. Pity those who don't like change.

JUST CAN'T GET ENOUGH SCIENCE

Much ink has been spilled using science for fictional productions, often called speculative fiction. For this reason, it seems pertinent to broach certain scientific topics not already introduced for clarification purposes. There are three I wish to address.

First, the initial way in which the idea of evolution was framed was descent with modification. This was the mood at the time of Erasmus Darwin, Charles' grandfather. The second came from Jean-Baptiste de Lamarck, when it was couched in terms of the inheritance of acquired characteristics. The claim was that whatever physical features underwent change based on need during the lifetime of an organism, these could be passed on to future generations, assuming the organism bearing it or them survived long enough to produce offspring. Even the early Charles Darwin held this view. Plus, contrary to popular conception, the later Charles self-identified as an agnostic, not an atheist, and perhaps surprisingly held to some version of ID. From the perspective of the twentieth-century biological community and onwards, Lamarck's approach was deemed heretical, since such changes do not become coded into the genetic material and are therefore not inherited, both Lamarck and Charles having preceded the genetic view of inheritance and thereby cannot be faulted for this.

It turns out that even now Lamarck is not so easy to get rid of, for the subfield known as epigenetics has unearthed that Lamarckian inheritance sometimes does in fact occur and the discipline has simply referred to it by another name. For example, in Sheldrake's terms, "the effects of toxins can echo for generations."[19] And further, "nutrition in childhood affect[s] the incidence of diabetes and heart disease in [the] grandchildren. Many common diseases that are inherited within families can also be passed on epigenetically."[20] Consequently, epigenetics does more than "[affect] which genes are 'switched on' or 'switched off,'"[21] although the expression will sometimes skip generations, thereby signifying a deviation from Lamarck's

---

19. Sheldrake, *Science Set Free*, 175.
20. Sheldrake, *Science Set Free*, 176.
21. Sheldrake, *Science Set Free*, 176.

idea. In any case, not all of which are deleterious, this marks another instance in which it would be hasty to charge a certain inconvenient theory with heresy.

And finally, on the topic of between which of nature versus nurture carries greater weight in, say, aggressive/violent/criminal behavior, a complex of genes may be involved, but the following analogy might prove helpful: nature ignites the car engine, while nurture puts it into gear and presses down on the accelerator. We can only hope that there is also a part of us which knows enough to apply the brake when needed.

The next subject to be treated is time travel—a favorite theme of many a novel and film, and those who avail themselves of it believe they have hard science in their corner. A figure no less than Albert Einstein is heralded as having championed it and his theories have supposedly proven it. More accurately, his ideas leave open the possibility of time travel, but mostly for subatomic particles, neither you nor me. To extend this into the human scale is very much the fictional aspect of the account, not least of which is for the following concern. What becomes problematic if we were to travel back in time, the more likely of the two directions, is that there is a complete host of objects that would need to come along for the ride.

To put it in the starkest possible terms, we would need to place all the pieces of the universe back into the coordinates they assumed at the specified instance on the time machine dials. This is the trouble when one lives in a relativistic cosmos where everything is relative to everything else. In actuality, the entire universe would need to reverse its tracks if time were to proceed backwards, for all of it becomes affected. Nothing in the universe currently occupies the location it did in the past, so the totality would need to be harnessed and relocated to its previous whereabouts if there were to be any going back in time at all. That's a lot of work and no terrestrial technology could accomplish it. At our level, then, it remains fictional.

Lastly, each particle has its associated field and vice versa. Space is something rather than nothing, and in it virtual subatomic particles continually surface on the order of nanoseconds and then, as mentioned, disappear back into the field, since the latter is not sufficiently strong so as to generate free-standing particles. Space is awash with fields and has an energy, as does matter in its own right, however the particles that are produced by the fields do not include hydrogen atoms or any other element in the periodic table, for space itself, as opposed to the objects in it, does not contain the raw materials to manufacture them. Thus the early Hoyle was in error.

Additionally, dark matter is the material believed to make up the missing matter of galaxies, the gravitational pull of which prevents galactic contents from flying off into space as mud from tires hits the mud flaps. The difficulty lies in that it does not interact in conventional ways with other matter, except gravitationally, meaning it evades detection. Further, dark energy is the energy of space and is a repulsive form of gravity. It gives rise to more space as a result of the universe's expansion. More precisely, as the universe expands it creates more space and dark energy fuels the expansion. Regrettably, dark energy, like dark matter, eludes observation.

In reference to Sheldrake and the manner in which he speaks of morphic fields, they seem to be well-nigh all-pervasive throughout the universe, and in these terms they are on a similar footing as dark matter and energy. One would suspect that if something were this pervasive it would leave some tracks, but his theories do not translate into practice and he receives grief for this. However, dark matter and energy do not make any tracks either, leaving the physical laws of the conservation of matter and energy in tatters. As Einstein informed us, matter/mass is energy, but contrary to him the amount of energy and therefore matter in the universe, as it happens, is increasing, owing to this dark energy and as opposed to these laws. And since none of morphic fields, dark matter, or energy can be detected, in Sheldrake's defense, if one criticizes one scheme, one should also criticize the other.

Also tossing the Scriptures into the mix, certain biblical scholars point to a document known as Q, from the German *Quelle*, meaning "source," to make sense of what they observe in the Synoptic Gospels. These academicians normally view the chronology of these three gospels as having come in the order of Mark first and then followed by Matthew and Luke, or for some of them Luke and then Matthew. The latter two each draw material from Mark and then also have other material in common. The question becomes: what was the origin of this source? Q is held to be the alleged work that both Matthew and Luke employed as a source in addition to what they derived from Mark. The trouble is that there is no such extant document, yet they insist that there must be one, otherwise how are we to explain the material common to both Matthew and Luke but absent from Mark? In the discipline of philosophy this is known as an appeal to ignorance—the temptation to posit that something must be the case, for what else could explain it? Well, the standard response could be, "use your imagination," for perhaps not all the available options have been exhausted.

Regrettably, some of those who are interested in the Q document have lost sight of the fact that while it remains undiscovered, it also remains a hypothetical text. To do otherwise commits another fallacy called an appeal to authority, for the tendency is to announce that many or maybe even most scholars consider Q to be real or that significant luminaries, such as the leaders in the field, find themselves amongst this august membership. Sadly, it does not matter who or how many reside in the Q camp, for that by itself does not ontologize the work, that is, catapult it from the ranks of the conjectural to the actual. Tangentially, it must further be stated that if so much zeal and fervor is directed toward a literary work that simply must exist, why is the same courtesy not extended to God? In either case, it seems that faith is being placed in something that or someone who cannot be established as definitive. If this is considered acceptable in science, it ought also to be so in religion.

Moral of the story, neither science nor religion is a stranger to theoretical entities: for religion, this Q document; for science, superstrings, the aforementioned dark matter and energy, plus idealized items such as a vacuum and frictionless surfaces are not so easy to locate in the universe.

TENTATIVE CONCLUSIONS

I find myself being driven to offer some additional comments to round out our discussion.

First, what we have is a failure to appreciate that most causes have effects and to foresee the long-range consequences of our actions. One technological advancement might yield eventual drawbacks. Freon is one such example. Its initial result was the saving of many lives by supplying a chemically safer way to refrigerate than ammonia plus sulphur dioxide, methyl chloride, and later methyl formate. But then it conversely endangered many lives by ripping holes in the ozone layer, which as of now has essentially been restored. One scientific improvement can lead to unexpected and unwanted outcomes. Bunnies are cute, so why not introduce them into a new and favorable environment to which they have not as yet been exposed? What could go wrong? Well, the rapidity with which they proliferate, for one.

The same is true for any delicate ecological balance. And as for that old bromide about whether a tree that falls in the forest makes a sound even if no one is present, well if a specimen of flora were to fell a promising and advantageous specimen of fauna on its way down, this could alter evolutionary pathways. The sound you would then hear could perhaps be that of a fragile species going extinct if its numbers have already dwindled. Our acts can recoil on us, so better exercise caution regarding the next scientific or technological breakthrough, for it could be our undoing, especially if throwing more technology at it is thought to be a corrective or should there be those around eager to corrupt it.

Moreover, another question worth asking is: Does natural selection know more and do more than it is letting on? What I mean is this. The human population has roughly equivalent parts of males and females. If you have a large enough sample size, this is the kind of result one is ordinarily going to find—half of one and half of the other. Nature usually works like this. Yet it is not the case with DNA. Certain biomacromolecules come in right- and left-handed varieties, about 50 percent of each, known as a racemic mixture. DNA, however, surfaces in only one of these types—right-handedness—100 percent of the time and zero of the other. What accounts for this?

That is why Francis Crick, famous for having teamed up with James Watson to uncover the structure of this very same DNA molecule in 1953, submits the notion of panspermia, which continues to have legs in some quarters. The idea is that Earth was seeded with one form of DNA from the planet Mars during the time when it exhibited life. DNA is easily denatured, but when carried on, say, an asteroid, it could make the trip as some amino acids also do. These latter organic molecules, though, normally display the kind of handedness that is not biologically active on our planet, often but not solely. Although a stretch on the face of it, there are few competing theories. This is the reason why Crick's hypothesis endures. But do keep in mind that in potentially solving how DNA arrived here, it thereby pushes back the origin question by asking: How then did it appear on Mars? Infinite regresses do not have many followers.

On a related issue, let's take an example from history. The bulk of humans engaged in war have predominantly been male, meaning by a war's end the male population of the participating nations has become somewhat depleted. Now of course worldwide this does not make a significant dent, but the numbers are appreciable and non-negligible. In subsequent years, inexplicably, there are a greater amount of male births in the affected countries so as to make up the difference. Our question then becomes how natural selection knows to do this, or is there another factor at work?

So too with global numbers. Until recently, there was concern that our projected population, given our rate of growth, would produce overpopulation before long. But should it be accurate that we are currently experiencing a negative growth in most countries, less than or barely reaching replacement levels, the danger might be avoidable. Is this a conscious decision on our part or is natural selection, thought to be a passive influence, acting in ways we have not come to expect? Rather than a detached sieve, is it or has it now become, perhaps for the first time, also paternalistic? If so, what brought about the change?

Even if overpopulation were not to be an issue any longer, where to bury those who are to come certainly is. Europe has dealt with limited space for a long time. Taking the German experience for instance; it has a policy of sharing plots and/or renting them, leaving the surviving family after a generation or so to relinquish the plot, should they elect not to renew, and have the new resident interred on top of the old. Nor are they permitted to have cremated remains in an urn in the home. Whatever the policies we ultimately adopt, real estate that could be used for other means, like agriculture or housing, may actually be dedicated to resting places for the dearly departed.

One possibility, I suppose, would be to use non-arable land, like deserts, for burial use. Anticipating an objection, some might not live close enough to one so as to visit the plots of friends or relatives with any frequency or at all. Yet we may need eventually to decide between visiting and eating. If tapping the conventional resources on land and in lakes and rivers, seas, and oceans has reached capacity, then we might be required to look elsewhere for victuals, or be innovative in employing other potential food sources not countenanced before, like insects and jellyfish. Here's hoping we develop a taste for the exotic.

Second, more on the brain, mind, and consciousness. In the formation of *Homo sapiens sapiens*, as intimated above, brains developed and grew but at one point diminished in size by ten percent. One question is whether we can be certain as to which part(s) of the brain were reduced or removed, and another is what if anything did the extra amount enable us to do for the time period we possessed it that we cannot do or at least do less of currently? At the same time, we cannot have complete knowledge of the brain since entertaining this very thought alters it and forms additional dendritic connections. It never sits still but keeps on changing.

The mind may best be described as a beacon, or even better yet, according to Sheldrake, both a transmitter and receiver, sending and accepting signals, as parapsychology might explain. Lastly, consciousness may not be confined to brains, and the part of the brain that activates and lights up might not be the cause of the event but its effect, meaning that consciousness could lie deeper, and thereby persons would be more complex than we anticipate. The brain is usually touted as the most complex thing we have found to date in the universe; personhood and all which this entails could outstrip it.

Third, if you are thinking of what the use may be in speaking about something as peripheral as parapsychology, despite the fact that it received, albeit reluctant, affiliate status in the AAAS (American Association for the Advancement of Science) in 1969 under the umbrella of psychology, well let's have another look at science. We live in a time when science by itself does not have the instant credibility it once enjoyed. Now it requires some assistance. There are those who believe that science has been co-opted by certain organizations having a personal financial stake in its results. Well, it has; this is not new and we *should* believe this. But what is different now is the lack of confidence in science when it has an agenda to uphold. We like science when it represents us as a spokesperson for our cherished beliefs, though distrust it when it testifies against them.

This is similar to the trust people once placed in television broadcasting, thinking it would not lie to us about anything important, would it?

Then came the scandal of a game show early on in TV history when a contestant was given access to the answers. At this point, TV fell from grace and we needed to develop filters when we viewed it. Science is experiencing this now, and not only in terms of the climate debate but also anti-vaccinators, and needs to press the reset button or run the risk of being labeled what much information is nowadays by those who dislike it, namely, fake news. Some are discounting its findings as telling just part of the story—the part they disagree with. Previously, science was not viewed as something one *could* disagree with, mainly because it was regarded as objective. These days, it is understood as containing biases like any other human endeavor. As a pendulum swing in the other direction, the knee-jerk reaction on the part of observers of science might be that it needs to earn its stripes of trust afresh.

There are times when science makes things difficult for itself when it comes to public relations. All that science studies is material reality and hence this is all it knows, but it is a huge leap to pontificate, as some scientists have, that this is all there is. Announcements such as reality being entirely defined by what we can observe (the empiricist perspective) are those of belief and not science, and they come from armchair philosophers who do not possess any more authority for wearing a lab coat. Those scientists who make such claims are out of their depth in this regard. Besides, there are things humans can experience which are not empirical, such as the aforementioned denizens of our subjectivity. What units does fear come in, for instance? I fear that scientists themselves have come across feelings such as these themselves. What scientifically do they make of them? That they are all in the mind? Even in suggesting this they are positing something non-material. Perhaps what they mean is that they are all in the brain. Well the brain is empirical, thus fear should be a scientific category in addition to the readiness for, say, fight or flight. We will eventually need to make up our brains/minds about this.

This physicality or substance view of reality is all that science has ever met up with, but this is a limited view and is not omni-competent. But since some are making this claim, or in scientific terms a hypothesis, then they will need to produce evidence for it, if anything could ever count as evidence. In this way, they are in the same boat as religionists. They might castigate religion for not being science, yet it does not bear any such requirement; though it could be conducted scientifically and, perhaps surprisingly, there is a journal of the same name (*Journal for the Scientific Study of Religion*).

As already alluded to, science has not always served itself well by the public. Let's take one of its branches, nutrition, as an example. Outcomes of studies from fish to nuts, chocolate to salt, tea to coffee, bottled water to alcohol, egg yolks to egg whites, butter versus margarine, red meat to (bringing physiologists into the fray) how much sleep we need, have alternated between pro and con. Vacillating between two opposing positions does not really mark advancement, so we really need to stay tuned for the latest studies.

Is it any wonder why some findings are met with skepticism? Some researchers make the appeal to "believe the science," but as with the climate example, science was once accorded reliability and ascribed credibility; now it needs to achieve it, meaning these appeals need to be made. The notion that one cannot argue with the science is ostensibly untrue. In reference to what to trust nutritionally, we would do well to consult multiple sources when compiling our grocery shopping list before we confer legitimacy on any of them, especially if the science has been non-independently underwritten by corporations having a stake in the outcome of the studies. In the meantime, the least we can say, lest we resign ourselves to avoiding any and all food items on this list altogether, is that the bottom line for many of these comestibles reveals an optimum level between too little and too much. Even an excess of water can become toxic for kidneys.

For our fourth comment we will need to begin by referring to a passage of Scripture at length:

> Micaiah continued, "Therefore hear the word of the LORD: I saw the LORD sitting on his throne with all the host of heaven standing around him on his right and on his left. And the LORD said, 'Who will lure Ahab into attacking Ramoth Gilead and going to his death there?' One suggested this, and another that. Finally, a spirit came forward, stood before the LORD and said, 'I will lure him.' 'By what means?' the LORD asked. 'I will go out and be a lying spirit in the mouths of all his prophets,' he said. 'You will succeed in luring him,' said the LORD. 'Go and do it.' So now the LORD has put a lying spirit in the mouths of all these prophets of yours. The LORD has decreed disaster for you." (1 Kings 22:19–23)

Now aside from the obvious difficulty classical theists run into with this passage, namely, how an omniscient divinity can fail to know which course of action would be the best one to take and what is in the "mind" of the spirits in God's heavenly court, consider the following. If we are to assume at least for the moment that this is the manner in which God and the court operate, then two issues surface.

Initially, God would not appear to be above the tactic of using deception to achieve God's intent; and second, does this episode imply that there is a range of gifts, talents, skills, and abilities which angels, like humans, can possess, making some of them more capable of and adept at a task than others? Some angels, it would seem, have better ideas than others, as advisors do in the service of a head of state. The more advisors, the more advice, and potentially the greater the confusion. Does God surround Godself with only the very best and most able and competent spirits or with many and varied ones?

Fifth, which language are we to suppose will be spoken, whether telepathically or otherwise, in the afterlife, such as one of the ancient languages comprising the Bible, namely, Hebrew, Aramaic, or Koine (common, popular) Greek? And if it were not to be English, then would we come with immediate facility in it or would it need to be learned? Conveniently, when angels are dispatched to convey messages from God, as far as the biblical accounts are concerned, they are typically in the language with which the recipients are familiar, otherwise they need not bother, since communication could not occur without an interpreter, and sometimes this too takes place (Daniel 5:22–28, the episode about the writing on the wall). So if the senders of these messages are sufficiently accommodating for us in this life, will we need to be the accommodating ones in the next, or will the transaction occur telepathically?—in which case no new language need be mastered on either side of this life.

Sixth and finally, as for the question as to why God did not begin with the new heaven and Earth at the outset as opposed to ending with it, I imagine one response could be that perhaps we need to go through boot camp first in order to prepare us for the life ahead. I am fully aware that some of the views above can be adjudicated as heretical from a traditionalist standpoint. Well, some of the best ones are! God has been pleased to use foolishness in its various forms to salutary ends.

DELIBERATE CONCLUSION

We have come a long way and covered a lot of ground, dealing with matters near and far in space-time together with the beyond. We have examined the vistas of three renditions of divinity on a scale from complete authoritarian rule to hardly any at all. We have further investigated humans and what they are like both here and now as well as what they might be like in a hereafter. We have considered the lifelong dendritic connections and reconnections together with their synaptic junctions as markers of personal identity and come to propose that many if not most of our nature's aspects can do the same and supply these fingerprints or signatures. We have also noted the capabilities as well as limitations of both religion and science. The conclusion is that we cannot be conclusive; issues remain speculative and conjectural, not definitive. But we need not leave it there. At this stage in my life and career, I am unencumbered by the fear of academic reprisals or any other repercussions and so am free to speak in line with my conscience.

In actuality I am an ally of both religion and science, but realize that both are in need of some renovation. While each of the topics we have covered might not instill confidence in pointing the way forward on their own, we are not restricted to their individual contributions, for collectively they may shed light on the yet unknown, illumine misty horizons and present a mosaic forming an image from which each piece becomes significant, even essential. Hence there is a place for both Plato and Sartre in theories of human nature; for all three perspectives on divine nature; for both Whitehead and Sheldrake to offer us a sense of the mechanics of what is just beyond our reach; for the biblical documents to serve as ancient cosmological wisdom; for both dualistic and integrated approaches to inform theological anthropology; for both reincarnation and resurrection to assist in theories of personal survival; for ESP as an undervalued source of knowledge; and all together ultimately bridging physics and metaphysics, religion and science. The key is the cumulative effect of all these elements and, as one can readily determine, the most salient method is to forge a path using a both-and instead of an either-or strategy.

An example of the biblical wisdom I am particularly fond of comes from Ecclesiastes, otherwise known as the Teacher. In one passage he observes that "the race is not to the swift or the battle to the strong . . . but time and chance [occur] to them all" (9:11). This is a scriptural way of saying that crap happens in human lives and has done so for as long as there have been humans. This is an acknowledgement that we live in a messy world, but the messiness should not stem from us. And as we mentioned in evolutionary theory, time and chance are not causes but arenas or occasions within which events can arise. Same thing here. Many and varied events have the potential of surfacing, even to those most well adapted. Further, it also suggests that the classical view of God as possessing and wielding complete rulership over all of nature lacks full biblical support. God might not be as much of a control freak as previously imagined.

And now a word about truth, which is not a popular topic these days. Some say it is outmoded and obsolete, as no one talks about it anymore; others that there simply isn't any. The work you have before you is largely about human nature (partly a scientific and partly a religious-philosophical concern) but not exclusively, for there were other pressing matters to consider as well, all of which, it is my hope, we would do well to take with us when we encounter this world of religion and science. In addition to prefaces and appendices, conclusions can permit authors to say what they really want to say. So allow me to talk about truth and the human condition.

We have all seen what humans are capable of: some good, some bad. As for the former, kudos; as for the latter, there is still time to alter our priorities. For now. There are times when I am proud of us, as when some race to the aid of others requiring disaster relief, and other times when I am disappointed—nay, ashamed—as in cases where corrupt officials manipulate the system of science and political policy to their own advantage and assure us that there is no proven health risk posed by substances like nicotine, GMOs, and the water in Flint, Michigan. Further, soil and water tend to become polluted in areas reserved for the proliferation of chemical companies. Big surprise. Toxic in, toxic out.

Anyone in a position of power—and most of us deploy some measure of it—can become corrupt. This includes institutionalized religion as well as industry-backed science, the type which has been underwritten by companies having a stake in the experimental findings. Neither by itself, due not least to its ambiguity, is salvific. We will need to look elsewhere. God help us in this. My hope is that we will be moved into action and propelled to meet the needs of others, and that this inclination will not wane once (or if and when) situations improve, but will become a tendency of which we will all be found guilty. Given the times we live in it seems appropriate to remind ourselves of the following. As for racial activism, if you are human, then you possess the union card allowing you to be here and share the planet, and so you should be extended this courtesy. And as for the pandemic: peace—pass it on; COVID—do not pass it on.

Two more items of note. In the first place, since there is no pure objectivity available to humans, all art, reporting, film, literature (including the Bible), and advertisement is propaganda, from the images on coins, currency, and stamps to the promise that the more money you have the happier you will be and the promissory note that science will eventually come to any and all rescues. This is because the people making them are revealing the world as they see it, from their perspective, and these are agenda-driven. They are urging us to see the world through their ineluctably, and not always recognized or acknowledged, limited eyesight. Marketing consultants will maintain that it's all in the packaging, so this is how messages are packaged. All agendas, in turn, are expressed in propagandistic terms. In the second, all social constructs and all political systems are broken. Why? Because the people in them are broken. This includes all of us. And that's the truth. But it's not the whole truth. And there is help. (Also recall that part in the short story about the love that endures beyond generations; well that's true too—it can outlast anything.)

In my estimation, the experiments of rugged individualism and the drive toward an autonomous self have been found wanting, and we live in a time when wisdom is in short supply. Remedially, my assessment is that, contrary to the opprobrium leveled against religion as either poisonous or an opiate of the masses, together with a safety net for those who cannot marshal enough strength on their own so as to cope and see their way through, that is, achieve self-sufficiency, pride does not prevent me from saying that I personally will accept all the assistance I am afforded. From all quarters. Even if it stems not only from the horizontal but also the vertical. The latter often works through the former, and we can all be facilitators in either this direct or indirect line of activity. And as for the parameters of science, we would do well to expand the borders of what constitutes legitimate science so as to cast a wider net for what is allowed to inform us, once again from all pertinent quarters.

Using terminology which the apostle Paul himself employed and antithetical to what Plato affirms, to see from behind a veil or through a glass darkly is language descriptive of what it is like to be, say, hidden in a second womb, to have moved from one womb to another in preparation for and in anticipation of awaiting another birthing event. For Plato, to be born is to incarcerate a soul and is therefore to be endured until death can release it. For Paul, likely to agree with the later gospel account, life is a gift though we yearn to see the "renewal of all things" (Matthew 19:28), necessitating analogously a breaking forth from underground through the planting of a seed and the eruption into a mature plant. We are first born from our mothers and lastly from the Earth. We return to the dust and then emerge from it. The Giver of Life blesses the first, having made it good, and the same agent in the role of Great Obstetrician orchestrates the last, making it permanent and perhaps even dust free.

> Looking forward and working while we wait (a deliberate alliteration),
> Signed, Herb Gruning

EPILOGUE

The Intrusion

Consider what lies below as an added bonus for those who have persevered with my reflections thus far.

THE INTRUSION

By all accounts I am a regular guy. What I mean is average. Physically I would describe myself as of medium build. I live an average life in an average city, I have an average home and drive an average car. Thankfully I also have average payments on an average mortgage. I have average attire, eat average meals and thus have an average diet, if there is such a thing, and have an average social life, though I never married—too average for comfort, I suspect, in the eyes of prospective mates. Since I have no kids, I have surrogate ones, so to speak, in owning both a dog and a cat. I introduced them to each other when they were a pup and a kitten, so they have become used to the idea of being around each other without creating a fuss. I suppose I am doing my bit for world peace, figuring that if two such mortal enemies can cohabit without the flying of any fur, then what is touted as a superior species ought to be able to do the same. So far the world has not taken notice.

Oh, before I forget, I also bear an average name—Stanley Abercrombie—and work at an average job as a clerk, who answers to somebody else, and who works for a company owned by somebody else. Don't get me wrong; I am not dissatisfied with my workplace, but have created pet names for my colleagues, many of whom are women, based on their not entirely unwarranted stereotypes. Miss Construe in accounting was always interpreting situations incorrectly. Miss Kreant could be counted on to do the wrong thing. Miss Sery was woeful about her failed relationships. Miss Sing was sometimes nowhere to be found, A.W.O.L. as the military calls them. Miss Nomer was often called by the wrong name, probably because she was frequently mistaken for a well-known starlet. And Mrs. Sippy, well, she was from the South.

Then there were the matronly types in archives. Auntie Jen was never sick a day in her life. Auntie Mony was an earthy sort. Aunt Ticks was known for her misadventures. Auntie Pasto was a culinary wiz, and Auntie Podal together with Auntie Thesis were always in opposition. Lastly, Auntie Kwarian was the eldest of the crew. There were also the brothers, Mr. Dearly and Mr. Deeply, who were heart-stricken about lost loves. The company was run by a married couple, Mr. Hitz and his Mrs., whose decisions made the firm's stock go up and down.

Now these pet names should not be taken as an indication that I cared little about them or wanted to distance myself from them. I just reckon it was the average thing to do. Indeed everything about me is average, which is why I was so surprised that I was singled out when the aliens landed. Allow me to explain.

It was a dark and gloomy night. No really. Even this tidbit was average. There was not a star to be seen and the noise of traffic drowned out any intrusion into our world. I had just come through an uneventful day and was experiencing a night without incident. (I am looking for ways to avoid using the term "average" unnecessarily.) When I think about events that might have led up to the intrusion and if anyone could have been credited as having been given advanced warning that an intrusion was imminent, nothing springs to mind, but I am certain that conspiracy theorists will have a field day. No world leader, at least, was offering any tip or giving a wink or a nod that we were about to be visited. Come to think of it, the visitors might have been preparing us and the rest of the world with our being completely unawares. This means we might never have known what to look for.

But back to this night. I would ordinarily have expected a huge ship to appear and hover over a major city and block out the sun, but this being a cloudy night, a ship would not have registered, at least to our unaided faculties. Nor was their craft of monumental proportions. Yet I am using ideas borrowed from Hollywood. They landed in the field behind my house. Nothing imposing about either them or their ride. Dare I say that they were average or that our expectations were unrealistic. But their very appearance here was unrealistic. I didn't know how to judge these events anymore. There I was, potentially witnessing perhaps the most extraordinary event in world history, and the first thing I thought to mention about them was where they parked. Admittedly, though, I was awestruck.

There were three of them—the aliens I mean. The one in the center seemed like the one in charge, and he (?) was flanked by two associates. They looked more human than some films about them made them out to be. Sure, the tops of their heads were larger than ours and their taste in clothes was questionable, but other than that they had at least external bilateral symmetry as do we along with the conventional appendages. I later learned their appearance was for our purposes, likely to circumvent the fear factor. The physical world seemed to be at their control, for they opened my front door without touching it and beckoned me to come outside to meet with them; otherwise I would have invited them in for a drink, an average one mind you. This would have been the neighborly thing to do, despite the intrusion.

Out of nervousness, I imagine, I exclaimed to them that I anticipated a much more grand entrance for their initial encounter with the world and us with them.

The one in charge said—although that is not quite the right way to put it, since his lips did not move, and I don't intend to imply that he was a ventriloquist:

[We are not here to impress] was what I seemed to experience him as communicating to me.

"Am I hearing your thoughts mentally?" I inquired.

[You will need to pardon us], he expressed, appearing to be annoyed, [it can be exasperating to explain the finer points of our existence and activities to an underdeveloped species.]

"So do I need to use my own voice or can I just think my thoughts?"

[You can do as you like; your intentions will get across.]

"Well, what have you come here for? Please tell us that it is not to be taken to our leader." I blurted this out mentally, just to see if it would work.

[We will permit questions once we have completed our mission. But since you asked, we have come to take with us a specimen of an average human for examination and training purposes. And by the looks of things, your world could use it.]

"Why would you select someone average when you could opt for the exceptional?"

[Because most people are average. Most cannot rise to the exceptional and we want to appeal to the planet as a whole.]

"And for those who are below average?"

[They will need to adapt or be left behind.]

"So if the world is mostly average, why am I chosen as opposed to billions of others?"

[Because you are not just the mean but also the median. Our statistics inform us that you are the midway point of all those currently on the planet. In essence, you excel at your mediocrity.]

I never looked at it that way before and felt a kind of reserved pride. "But what is wrong with our planet and in what way are we underdeveloped?"

[Compared to our civilization, you as a species should have moved well beyond this point by now.]

And with that he went on to explain their advanced society. Here are some of the details, certain items of which were couched in a condescending fashion.

Their own planet could have gone in one of several ways, most of them destructive, were it not for a global decision to combine their efforts to do all they could to survive both as a species and a world. They elected to relinquish any and all sorts of rank, status, and wealth so as not to lay claim to anything. It came down to a matter of doing this or dying. They communally took ownership of the planet by not taking individual ownership of anything. That way, all efforts at individual security were yielded so that there could be more collective security. By denying themselves any and all entitlements, this entitled them to work for a common cause. That in turn expedited the evolutionary process, given that less effort needed to be devoted to defense, and they became stronger mentally. By cooperating they were able to extend their lineage. They had no need for incidentals and hence could concentrate on the technology that would gain them the ability to leave their world once it became uninhabitable due to a dying star. They could then seek other worlds more compatible for them, at least for the time being.

They are now the generation which has made the fourth planetary departure and relocation in their civilization's history. They mentioned that it can be exhilarating to land on a planet for the first time in preparation for making it their home. They also informed me that our own lineage is in danger from a rapid increase in those genetic conditions giving rise to what we know as autism and other like challenges, in addition to nations being suspicious of other nations. In the short term, they warned, energies will be needlessly directed toward national rather than global issues, and much resources will be allocated for an ever-increasingly challenged population. In the long term, suspicion will escalate into outright intra-species aggression of nation against nation, as well as a diminution of individuals having the capacity to lead. I must admit that I was intrigued as to how they managed to survive when we might not, and if their strategy could work for us and our planet.

They took me aboard their craft, which I found to be remarkably spacious given its size. But I sound like a used car salesman. I wanted to ask them if it gets good mileage, but I resisted the temptation, seeing as I did not want to find out what their code of etiquette would unleash toward those who are flippant. I did, however, feel brave enough to ask them their age. They replied that they did not measure time in the same way we do, since it is not characterized by the same astronomical markers as ours. Their planet has a different diurnal period as does its voyage around their star, hence both the rotation and revolution differ. The least they could say is that they age more slowly and live longer than their earthly terrestrial counterparts. I then asked if their population had the same number of genders as ours, to which they responded that their, what we call women, actually operate their craft, for the purely logistical reason that they have a greater life span than males and interstellar travel is time consuming.

"I know what you mean," I intoned, "I can hardly endure sitting for eight hours on a plane."

[There is absolutely no comparison], they flatly retorted.

"What then are the tasks of males?"

[We are engineers and technologists], they chimed, [while the women are navigators and scientists.]

I restrained myself from asking which gender would be the most willing to stop and ask for directions, for I could tell that such impertinence was becoming tiresome for them.

"One more thing," I ventured, "do you reproduce in the usual way?"

[We take a dim view of inquiries such as this. Of course we have no queen as do your ants.]

"Fine. What then do you regard as your mission here?"

[We are about to embark on our fifth relocation in the galaxy, as our current planet is nearing its expiration, as will yours, from our star's next stage as a red giant. We are scouting other locations for habitability. Yours is set to last for another five to six billion of your earth years, which is suitable for us.]

"Hmm, just at the time when we are considering terraforming Mars."

[That we find to be premature, for your world has more in favor of it than a move elsewhere.]

"So let me get this straight. You have been looking over our planet as a realtor would properties."

[Yes, and have done so for a long time. We needed to make a thorough search before reaching the correct decision. And we have made it. You should feel honored.]

"But this is our place, making you intruders."

[Recall what we mentioned about laying claim to things. You do not own your planet; you merely find yourself on it. And we expected resistance. This is why we selected you to break the news to your world. Coming from an average person who is onboard with our program, we could reach the most people.]

"You have underestimated our citizenry. We are not likely to give up what we have so willingly."

[We have studied your species and its bellicose nature. We do not insist on compliance, which is why we need to make appeals. If we can convince you, perhaps we can reach others.]

"What stops you from simply taking what you want?"

[We are not like you. We do not impose our will but attempt to convince using reason. We have lived on and departed from four different planets to this point. We recognize a responsibility to manage the galaxy well; not only our planets but those beyond. Yours frankly needs it.]

"I can see that ultimatum being a tough sell to those who have emotional investment in what they call their own. We are unlikely to share, not least due to the fact that there are already enough of us and we are close to reaching sustainable capacity. How many of you are there anyway?"

[Many would be a good description. We would, as you call it, terraform your deserts and oceans where few reside. In return, we would require you to adopt the program to relinquish ownership of possessions so that all might not only survive but flourish and thrive. And kindly refrain from depending on the drivel one of your authors has composed about our succumbing to a terrestrial microbial agent. We are immune.]

"Well, if I am not convinced, then I doubt I could convince others."

[This is what we feared.] I could tell that things were getting a little heated. [Why is your kind so stubborn as not to see what is plainly best for it?]

"Just unlucky, I guess. You chanced upon a species that likes to call its own shots."

[We were right; you are underdeveloped and that could be your undoing.]

"We would rather strike out on our own than be struck by someone else."

[What you view as struck is much more a cooperation. What prevents you from doing this?]

"I suspect that we will say, 'Only if it's on our own terms.'"

[We can see that we are getting nowhere and that our efforts will come to naught. You do not even deserve our assistance.]

"I imagine we will only accept assistance if we ask for it."

[We are offering you survival; you need only adopt our program.]

"That would sound too communistic for us and we would only see red, literally. I'm afraid you have picked the wrong species."

[Then we have nothing further to discuss. We will take our efforts elsewhere and leave you to your own devices.]

And with that they summarily ushered me out of their craft, somewhat rudely, I thought. Before they left, I asked them, "Wait, what are your names?"

They merely blurted out, [Our forms would be unrecognizable to you and our names unpronounceable. It seems little of our kind corresponds to yours.]

With their departure, I was feeling self-congratulatory, thinking I had just saved our world. Mr. Average has won. Exactly what, though, I was unsure of. I returned home only to find that my front door was closed and locked.

BIBLIOGRAPHY

Allison, Dale C., Jr. *The Historical Christ and the Theological Jesus*. Grand Rapids: Eerdmans, 2000.

Barbour, Ian G. *Religion and Science: Historical and Contemporary Issues*. San Francisco: HarperSanFrancisco, 1997.

———. *When Science Meets Religion: Enemies, Strangers, or Partners?* San Francisco: HarperSanFrancisco, 2000.

Berman, Morris. *Coming to Our Senses: Body and Spirit in the Hidden History of the West*. Toronto: Bantam, 1990.

Brockman, John, ed. *The Universe: Leading Scientists Explore the Origin, Mysteries, and Future of the Cosmos*. Toronto: Harper Perennial, 2014.

Carter, Chris. *Science and the Near-Death Experience: How Consciousness Survives Death*. Toronto: Inner Traditions, 2010.

Clancy, Susan A. *Abducted: How People Come to Believe They Were Kidnapped by Aliens*. Cambridge, MA: Harvard University Press, 2005.

Dehaene, Stanislas. *Consciousness and the Brain: Deciphering How the Brain Codes Our Thoughts*. New York: Penguin, 2014.

Goodacre, Mark. *The Case against Q*. Harrisburg, PA: Trinity, 2002.

Gould, Stephen Jay. *Full House: The Spread of Excellence from Plato to Darwin*. New York: Three Rivers, 1996.

Griffin, David Ray. "Of Minds and Molecules: Postmodern Medicine in a Psychosomatic Universe." In *The Reenchantment of Science: Postmodern Proposals*, edited by David Ray Griffin, 141–63. Albany, NY: State University of New York Press, 1988.

Gruning, Herb. *God and the New Metaphysics*. Nevada City, CA: Blue Dolphin, 2005.

———. *God Only Knows: Piecing Together the Divine Puzzle*. Nevada City, CA: Blue Dolphin, 2009

———. *How in the World Does God Act?* Lanham, MD: University Press of America, 2000.

———. *Who Do We Think We Are?: What It Takes to Be Human*. Nevada City, CA: Blue Dolphin, 2015.

Hartshorne, Charles, and William L. Reese. *Philosophers Speak of God*. Toronto: University of Toronto Press, 1963.

Heine, Steven J. *DNA Is Not Destiny: The Remarkable, Completely Misunderstood Relationship between You and Your Genes*. New York: Norton, 2017.

Kaplan, Matt. *Science of the Magical: From the Holy Grail to Love Potions to Superpowers*. Toronto: Scribner, 2015.

Kuhn, Thomas S. *The Structure of Scientific Revolutions*. 2nd ed. Chicago: University of Chicago Press, 1970.

Lewis, C. S. *The Four Loves*. London: Fontana, 1963.

MacIntyre, Alasdair. "The Story-Telling Animal." In *Twenty Questions: An Introduction to Philosophy*, edited by G. Lee Bowie, Meredith W. Michaels, and Robert C. Solomon, 368–73. 3rd ed. Toronto: Harcourt Brace, 1996.

McDonald, Bob. "Interview with Brother Guy." *Quirks & Quarks*, CBC Radio (London, Ontario 93.5 FM), May 20, 2013.

Nichols, Terence. *Death and the Afterlife: A Theological Introduction*. Grand Rapids: Brazos, 2010.

Pickover, Clifford A. *The Paradox of God and the Science of Omniscience*. New York: Palgrave, 2001.

Plato. *The Republic of Plato*. Translated by Francis MacDonald Cornford. 1941. Reprint, New York: Oxford University Press, 1970.

Schellenberg, J. L. *Evolutionary Religion*. New York: Oxford University Press, 2013.

Sheldrake, Rupert. *Dogs That Know When Their Owners Are Coming Home: And Other Unexplained Powers of Animals*. New York: Three Rivers, 1999.

———. *A New Science of Life: The Hypothesis of Formative Causation*. London: Paladin, 1987.

———. *The Presence of the Past: Morphic Resonance and the Habits of Nature*. Rochester, VT: Park Street, 1995.

———. *The Rebirth of Nature: The Greening of Science and God*. Rochester, VT: Park Street, 1994.

———. *Science Set Free: Ten Paths to New Discovery*. New York: Deepak Chopra, 2012.

———. *The Sense of Being Stared At: And Other Aspects of the Extended Mind*. New York: Crown, 2003.

———. *Seven Experiments That Could Change the World: A Do-It-Yourself Guide to Revolutionary Science*. New York: Riverhead, 1995.

Sheldrake, Rupert, and Michael Shermer. *Arguing Science: A Dialogue on the Future of Science and Spirit*. Rhinebeck, NY: Monkfish, 2016.

Smolin, Lee. *Time Reborn: From the Crisis in Physics to the Future of the Universe*. Toronto: Viking Canada, 2013.

Stevenson, Ian. *Where Reincarnation and Biology Intersect*. Westport, CT: Praeger, 1997.

Sweeley, John W. *Reincarnation for Christians: Evidence from Early Christian and Jewish Mystical Traditions*. Nevada City, CA: Blue Dolphin, 2013.

Weinstein, Larry, director. *Propaganda: The Art of Selling Lies*. Hawkeye Pictures, 2019.

Whitehead, Alfred North. *Process and Reality: An Essay in Cosmology*. Corrected edition. Edited by David Ray Griffin and Donald W. Sherburne. New York: Free Press, 1978.

———. *Science and the Modern World*. Lowell Lectures, 1925. New York: Free Press, 1967.

www.ingramcontent.com/pod-product-compliance
Lightning Source LLC
Chambersburg PA
CBHW050809160426
43192CB00010B/1696